# Atypical Deglutition Treatment

Antonio Amitrano · Francesco Benso

# Atypical Deglutition Treatment

## The Role of the Executive Function

Antonio Amitrano
Speech Therapist
Rome, Italy

Francesco Benso
Albenga, Italy

ISBN 978-3-032-01484-9     ISBN 978-3-032-01482-5   (eBook)
https://doi.org/10.1007/978-3-032-01482-5

The original submitted manuscript has been translated into English. The translation was done using artificial intelligence. A subsequent revision was performed by the author(s) to further refine the work and to ensure that the translation is appropriate concerning content and scientific correctness. It may, however, read stylistically different from a conventional translation.

Translation from the Italian language edition: "Rieducare la deglutizione atipica" by Antonio Amitrano and Francesco Benso, © Carocci Editore 2022. Published by Carocci Editore. All Rights Reserved.



This Springer imprint is published by the registered company Springer Nature Switzerland AG
The registered company address is: Gewerbestrasse 11, 6330 Cham, Switzerland

If disposing of this product, please recycle the paper.

*Whatever you do may not make any
difference, but it's very important that
you do it.*

*Gandhi (2019)*

*Keep planting your seeds because you won't
know which ones will grow, maybe they
all will.*

*Einstein*

# Foreword

In recent decades, the central role of the tongue in shaping the morphology of the stomatognathic system and all related apparatuses has emerged with increasing evidence. It has been understood that atypical swallowing, essentially characterized by the failure to transition from an infantile to an adult swallowing pattern, is responsible for pathological modifications affecting a wide range of oro-facial functions: dental and dento-skeletal malocclusions, alterations of the correct nasal respiratory functionality with severe inflammatory consequences affecting the lymphatic structures of the oro-pharynx, speech deficits, and dysfunction of the Eustachian tube that expose the child to recurrent or chronic catarrhal otitis.

In other words, it was understood that it is not the form that determines the function... but it is the function that determines the form, and all the professionals involved in the care of oral pathologies—speech pathologists, orthognathodontists, otolaryngologists, and Foniatristi—now have to recognize in the re-education of deviant lingual functionality the keystone to make their therapeutic approaches effective. All this would explain the extraordinary spread and refinement of myofunctional techniques used by speech pathologist to make young patients acquire the correct swallowing model.

In a literary and operational panorama characterized by well-established rehabilitative proposals, focused on myofunctional re-education, this manual breaks in, whose authors suggest a new rehabilitative approach, no longer centered on muscle re-education, but on a top-down learning of the proper motor behavior, based on the motor learning principles.

This study offers the opportunity to trace to the smallest detail the development of swallowing in childhood and its evolution in the different stages of life. It also allows for an in-depth review of the principles of motor learning and the possibilities of applying them to the learning of the correct lingual movement.

Despite clinical effectiveness studies still being in progress—with the first scientific study expected to be published shortly—the Amitrano Benso Executive Attention and Swallowing (ABAED®) method is a proposal of particular interest, considering the success that techniques of both motor learning and enhanced executive attention have already achieved in facilitating learning and speeding up progress.

The experimentation of clinical effectiveness still in progress—the publication of the first scientific study is expected shortly—have already achieved in facilitating learning and speeding up the acquisition of skills in a large number of other fields of application.

Phoniatric-University of Rome "La Sapienza"                    Giovanni Ruoppolo
Rome, Italy

# Introduction

"Bare life is breath and nourishment": with this phrase, theologian Vito Mancuso (2015) describes the reality of human existence, which needs air and food to continue to exist. Eating and breathing are the activities that make a living being an entity, and at the same time they express the two needs that no being can escape, at the risk of the end of biological survival itself. Over the course of evolution, specific anatomical–functional systems have been structured at the body level, which allow man to bring oxygen into the lungs and food into the digestive system. The respiratory and nutritional/digestive systems share, at least in their first part, the same anatomical structures which, under normal conditions, through a complex neuromuscular system, allow oxygen to enter the lungs and food to enter the stomach, preventing dangerous deviations along their way. In fact, the accidental entry of bolus into the airways can cause death by suffocation. Swallowing is the function that leads food into the stomach. Under normal conditions, it occurs simultaneously with breathing, while a complex valve system prevents air from ending up in the stomach and food in the lungs. Breathing and swallowing are functions that the newborn is able to use at the moment of birth, while it takes about a year to learn to walk and a longer period to pronounce the first words. The first wail is the audible sign of the eruption of air into the lungs. At the first moments of life, the newborn approaches the mother's breast to begin suckling milk. The mechanism of swallowing is already present during intrauterine life; between the tenth and eleventh week of gestation, the first swallowing acts are already evident. Swallowing activity at this stage plays a role in the management of amniotic fluid, but it primarily represents a training for mechanisms that at birth allow the newborn to feed and survive. Swallowing accompanies the existence of the individual from the first to the last moment of life. In fact, in addition to its importance for food intake during meals, it ensures the management of saliva day and night. The constant presence of swallowing in the individual's life undergoes a variety of modifications in its mechanisms. These changes are linked to both the morphological modifications occurring, in the different stages of life, within the apparatuses that allow swallowing—first of all the mouth—and to the variations in the kind of foods consumed by the individual. In fact, the newborn feeds exclusively on mother's milk or, in its absence, a substitute that, in any case, is of liquid consistency. With the weaning, he progressively begins to have foods of greater consistency. Milk is sucked from the mother's breast and/or from the nipple. The specific anatomy of the oro-pharyngo-laryngeal tract allows the newborn to feed safely and grow until the next step, when the diet is enriched with non-exclusively liquid foods. In this phase, the food approaches the mouth through different tools and no longer just from the mother's breast or the nipple. With growth, the mouth undergoes modifications as well as the pharyngo-esophageal tract, where the larynx deepens in the neck, increasing thus the phonatory possibilities of the individual, but exposing at a higher risk the bolus transit in the pharyngeal phase. During the course of life, all the phases in which swallowing is traditionally divided undergo modifications, particularly evident in the oral phase, which is constantly reshaping as a result of both structural changes, such as tooth eruption, and contingent conditions, e.g., a toothache or lingual aphthous ulcerations. In old age, swallowing undergoes a regressive process also linked to modifications of the swallowing organs, such as the loss of dental elements.

Two key moments can be highlighted in the continuum of swallowing modifications: the transition from suckling to sucking (the two different modes of neonatal suction) and the transition from infantile to adult swallowing. The transition from suckling to sucking is connected to the introduction of non-liquid foods and to the maturation of the cortical centers. In fact, in this phase, the exclusive control exerted by the brainstem evolves into a mixed form of control of the swallowing movements, by both cortex and brainstem, which makes swallowing no longer a completely reflexive process, with the first appearance of voluntary phases. The coexistence of voluntary movements and reflex activity characterizes the swallowing of the adult in which the oral phase is largely under the control of the cortex, while the pharyngeal and esophageal phases are almost entirely under the control of the brainstem. The second major moment of transformation in the ontogenetic evolution of swallowing is constituted by the transition from infantile to adult swallowing. In this case, the shift is slower and more gradual and, unlike the one from suckling to sucking, it doesn't exclusively depend on the maturation of the cortical centers but is also linked to the variation of the morphology of the mouth. Thanks also to the increase in the internal diameters of the mouth, as well as the transition to complementary feeding, the child begins to use—at first sporadically and later in more consistent ways—an up-down tongue movement pattern (on the vertical plane) that will gradually replace the in-out pattern (on the horizontal plane) typical of infantile swallowing. In other words, the larger dimensions of the mouth allow the tongue to move in more varied and wider directions. The change from infantile to adult swallowing generally occurs between the first and seventh year of age; broadly speaking, this process shall be considered as concluded with the eruption of permanent dentition. Rather than increasing a child's feeding capacity, the transformation of swallowing patterns will primarily affect the proper relationship between mouth structures, in general, and the bite, in particular, and the overall postural balance. It is worth underlining the close relationship between the tongue's posture and dynamic behavior and the harmonious development of the spine. The transition from infant to adult swallowing occurs in an evolutionary window that begins roughly with the first intake of solid foods and ends with the completion of the eruption of permanent teeth. In some children, for reasons that are not yet understood, this transition does not occur, leading to the persistence of an infant swallowing mode in later ages. This phenomenon, known as "atypical, deviated, or dysfunctional swallowing," involves a series of dysfunctions that need to be prevented or corrected by acquiring the adult swallowing mode. The commitment of the speech therapy world in this area has been considerable, leading to the flourishing of important and widespread rehabilitative methods. This book is not just a proposal for a solution to atypical swallowing; it aims to change the perspective on how to enable adult swallowing. It proposes replacing a bottom-up with an up-down approach. In other words, we do not start from a muscular perspective, but from a cognitive one, according to the assumption that every motor act and action within our body is planned, directed, controlled, and driven by the brain. Swallowing is no exception to this rule, as it is—especially in the oral phase—a voluntary act under the direct control of the cerebral cortex. This book will present the ABAED® method (Amitrano Benso Executive Attention and Swallowing), a rehabilitative, or, in better words, an "enabling" methodology aimed at facilitating the transition from infant to adult swallowing. To acquire the retention and transfer of motor learnings, this approach will rely on both knowledge about *motor learning* and the role of executive functions in motor learning itself.

This method has an empirical genesis, although it incorporates principles validated by research in the field of motor learning. Clinical experience gained over years of use has shown its relative simplicity of application and wide acceptance by subjects of various ages. While the study has not reached a final stage and still requires further steps of scientific validation, in our opinion, the time has come for its presentation to the general public.

In Chap. 1, the modifications of swallowing during an individual's life are described. It analyzes the physiology of swallowing in prenatal, neonatal, and infantile age, as well as in adulthood and finally in senescence. The modifications of swallowing are determined by anatomical and functional variations, as well as by different environmental stimuli at various times of life. Chapter 2 outlines the basic notions of attentive networks and working memory, as well as the possible interactions of these networks with motivational and emotional systems. It also illustrates the executive attentive bases of motor learning and the varieties of how dual tasks are used.

Chapter 3 is entirely dedicated to atypical swallowing and presents the ABAED® method, which the authors propose as a (re)habilitative modality for it. The authors reiterate that one of the method's unique features is its replacement of a bottom-up approach with an up-down one, which was developed based on the most recent knowledge in the field of cognitive sciences.

This book is dedicated to those who, not satisfied with their own knowledge, look forward with curiosity and openness to new ideas. It is also dedicated to the many people from whom the authors have learned and to the patients with whom and for whom we have been working.

The authors thank Francesca Danini, Giulia Rossi, Floriana Failli, Mauro Bonazzi, and Nicoletta Benso for their patient and learned collaboration.

Antonio Amitrano
Francesco Benso

# Contents

# Swallowing: Definition and Evolution in the Different Stages of Life

**1**

## 1.1 Swallowing in the Stages of Life

Eating is one of the two vital functions for survival. "A baby girl is born and the first thing she does is breathe, the second is trying to feed herself. As her whimpers show, she knows nothing and wants nothing else. Bare life is breath and food…She will need someone to support her for walking, she will need other human beings for learning to speak, but she doesn't need anything for breathing and eating: she already knows how to do it herself" (Mancuso 2015). Eating is not only essential for biological survival, but it also plays a fundamental role in the life of every human being for the many meanings it has as far as the relational, affective, and religious aspects of life are concerned (Amitrano 2024). The set of values of food has a strong impact in determining quality of life since, as mentioned, food is not only the energy support of the body but also an opportunity for participation in social life. It should also be considered that food is a source of pleasure for all the senses, not only, obviously, for taste and smell but also for sight and touch: In star-rated cooking, plating is a determining element for the success of a dish. The direct contact with the food that you get by eating with your hands has contributed to the success of finger food, in which handling with fingers represents a key component in the consumption of certain foods.

Humans, and not only, carry nutritional elements from the lips to the stomach through swallowing, a function that has been developing and changing over the course of phylogenetic evolution and that follows a specific path of changes in the ontogenesis of individuals. Swallowing should then be considered as:

*Thanks to Francesco Benso, a great beacon of knowledge and humanity, my teacher and friend. Thanks to Maurizio Cannavò who granted me the privilege of being his friend and being able to follow him in his work.*


A. Amitrano, F. Benso, *Atypical Deglutition Treatment*,
https://doi.org/10.1007/978-3-032-01482-5_1

A primarily complex, dynamic feeding skill, with macroevolutions (from neonatal swallowing to adult swallowing) and microevolutions (according to the changing eating habits of the in-group, for example, Asians versus Westerners, or of the individual, in relation both to the variability of traditional practice and physiological or pathological food needs); closely connected with complex servosystems such as, among others, salivary secretion; linked to the interface between the individual anterobuccal environment of food intake (suction or bite) and its possible oral processing (Schindler et al. 2011).

When you want to explain something and even more when you intend to modify it, you must first of all understand it (Parisi 2021). This is the right time to open a wide parenthesis on the ways in which swallowing takes place at different ages of life.

Swallowing is a complex act that involves about 25 pairs of muscles in the oro-pharyngo-laryngo-esophageal district and 5 pairs of cranial nerves (Dodds et al. 1990), as well as the brainstem, the cerebral cortex, and a number of cortical and subcortical areas (Jean 2001; Hamdy 1999; Hamdy et al. 1996; Mosier et al. 1999). The complexity of the swallowing act is further determined by its close correlation between breathing and phonation, functions that for the most part share the same anatomical structures. Swallowing occurs with a significant involvement of the central (CNS) and peripheral nervous system both on the receptive and efferent side, with the activation of the cranial nerves, the brainstem, the cortex, subcortical centers, the limbic cortex, and the neocortex. Some phases of swallowing are merely reflexive but this does not represent the totality of swallowing, which has a significant voluntary component. Every single swallowing act lasts about a second, and in adults, it is repeated on average a thousand times within 24 h (Dodds 1989). Swallowing occurs day and night, when its frequency is reduced to approximately 2–6 acts/h, while during lunch, it increases up to 300/h and stabilizes at about 10/h in the postprandial period (Martin et al. 1994b). Swallowing not only transports into the stomach foods of liquid, creamy, solid, gaseous, and mixed consistency, but it also carries to the digestive tract the secretions of the upper airways and the substances refluxed from the stomach itself. Swallowing also plays an important role in determining the air pressure inside the middle ear. The tympanic cavity is anatomically connected to the rhinopharynx through the Eustachian tubes (Passali 1985, 1989), anatomical osteocartilaginous ducts with the function of ventilating and protecting the middle ear and drain it from any possible exudate (Morgon et al. 1985; Coppo et al. 1966; Riu et al. 1966; Passali 1989). Through the Eustachian tubes, the air comes and goes to the middle ear and from here to the inner surface of the tympanic membrane with the role of balancing the pressure exerted on the external surface of the tympanic membrane through the air contained in the acoustic meatus (Serra 2011). This ensures a state of isobaricity on the internal and external surface of the tympanic membrane, which is a key condition for the correct acoustic transmission. The Eustachian tubes are permanently closed, which actively or passively open only under specific conditions. One of these coincides with the passage from the oral to the pharyngeal phase of swallowing, when the tubes actively open during the lifting of the palatal veil due to the contraction of the peristaphylin muscles. The way in which the oral phase of swallowing occurs affects the mechanisms of tube opening and therefore the efficiency of the functions they perform. Tongue thrust swallowing negatively affects the functioning of the tubes (Jonas et al. 1978) whose dysfunction, in the end, negatively affects language and the child's school performance (Rosenfeld et al. 2016).

Swallowing, while conveying food from the mouth to the stomach, at the same time protects the lower airways from accidental entry of parts of the bolus, thanks to a sophisticated set of safety mechanisms, the number of which underlines the relevance of the risks, to which the very survival of those who eat is exposed. The first of these mechanisms is the upward and forward displacement of the larynx during the pharyngeal phase, which prevents the larynx from being directly exposed to the trajectory of the bolus in its progression toward the stomach. The overturning of the epiglottis, moreover, diverts the bolus toward the piriform sinuses and the esophagus while simultaneously closing, although not completely, access to the entrance of the larynx. At the same time, the arytenoids move medially and forward toward the petiole of the epiglottis, while the vocal cords close, although with a delay of 0.6 s compared to the movement of the arytenoids (Kendall and Louie 2003). Another protective mechanism to consider is the way in which swallowing and breathing are coordinated. The act of swallowing generally occurs within the phase of exhalation so that in healthy subjects, inhalation is followed by a first phase of exhalation which is interrupted to allow swallowing and completed by the entry of the bolus into the esophagus (Martin-Harris et al. 2003; Palmer and Hiiemae 2003). This kind of coordination is by no means a protective system because any residue of the bolus still present in the oropharyngeal cavity will lie in an outgoing current driving it outward according to the subject's age and in case of pathological conditions (Le Jemtel et al. 2007; Mokhlesi et al. 2002; Zheng et al. 2016; Good-Fratturelli et al. 2000; Shaker et al. 1992). Another protective mechanism is cough, a paroxysmal expiratory act, evoked reflexively when a particle of bolus enters the airways. Cough has the function of further enhancing the defense mechanisms while, on the contrary, its weakness increases the risk of aspiration pneumonia (Ebihara et al. 2012). All these safety and protection mechanisms of the lower airways are implemented in the pharyngeal phase, which lasts about 0.7 s (Kendall and Louie 2003).

The accidental or disease-related malfunction of the protection mechanisms determines situations with possibly tragic outcomes. Every year in our country, about 300 cases of foreign body obstruction are reported in children under 14 years of age. Among children from 0 to 4 years old, choking is the second cause of death after road accidents; unfortunately, about 50 children die each year from a bite gone sideways, almost one child per week. Even higher is the number of deaths from aspiration pneumonia, a distinct subtype of pneumonia resulting from the passage of bolus into the lower airways (Marik 2001, 2011; Manabe et al. 2015).

Swallowing is not a static process; it changes in the various ages of life due to the modification in the anatomy of the swallowing tract, to the variation in texture of the bolus, and to the different ways of taking the bolus itself. In the evolutionary phase, significant anatomical changes occur, mainly due to the growth of the organs involved in swallowing: The palatal veil changes its dimensions and becomes more mobile; the pharynx lengthens and widens; the larynx grows and changes its position in the neck. Overall, the entire facial skeleton changes its size by expanding downwards and forward. The oral cavity also expands its internal diameters, allowing the tongue to increase its volume and to move more freely, thus expressing a more varied repertoire of movements. Within the mouth, the gradual eruption of the deciduous dentition takes place as well, the so-called "milk teeth," so called because they are destined to be replaced by permanent teeth. Milk teeth are generally smaller and lighter than permanent teeth and, unlike these, they have less pronounced cusps (Table 1.1). The two dentitions also differ in the number and type of teeth, with the deciduous dentition consisting of 20 teeth: 8 incisors, 4 canines, 8 molars, and no premolars. The permanent dentition, on the other hand, consists of 32 teeth: 8 incisors, 4 canines, 8 premolars, and 12 molars (Table 1.2).

With growth, along with the entire larynx, the epiglottis, which in the newborn is in contact with the palatal veil, progressively descends into the neck. Infants and small children up to about a year and a half continue to have a high larynx that extends approximately from the first cervical vertebra (C1) to the upper margin of the third or fourth cervical vertebra (C3–C4) in newborns and descends slightly to level C2–C5 in children of about 2 years of age. The high position of the larynx, with the separation of the airways from the digestive ones, allows the child to breathe and swallow some liquids almost simultaneously; in fact, newborns are able to breathe through the nose while sucking milk. The descent of the larynx into the neck, in a position between the upper edge of C3 and the lower edge of C5 in a 7-year-old child and between the lower edge of C3 or upper part of C4 and the upper edge of C7 in an adult, is one of the characteristic aspects of human ontogenesis. The change in position of the larynx in the neck profoundly modifies the patterns of breathing and swallowing, and while on the one hand it amplifies the possibilities of articulation of spoken language, on the other hand, it actually makes the transit of the bolus more at risk as a result of the anatomical intersection between the respiratory and swallowing functions.

**Table 1.1**  Milk or deciduous teeth

| *Upper teeth* | *Erupt* | *Fallout* |
|---|---|---|
| Central incisor | 8–12 months | 6–7 years |
| Lateral incisor | 8–13 months | 7–8 years |
| Canines | 16–22 months | 10–12 years |
| First molar | 13–19 months | 9–11 years |
| Second molar | 25–33 months | 10–12 years |
| *Lower teeth* | *Erupt* | *Fallout* |
| Central incisor | 6–10 months | 6–7 years |
| Lateral incisor | 10–16 months | 7–8 years |
| Canines | 17–23 months | 9–12 years |
| First molar | 14–18 months | 9–11 years |
| Second molar | 23–31 months | 10–12 years |

**Table 1.2**  Permanent teeth

| Upper teeth | Erupt |
| --- | --- |
| Central incisor | 7–8 years |
| Lateral incisor | 8–9 years |
| Canines | 11–12 years |
| First premolar | 10–11 years |
| Second premolar | 10–12 years |
| First molar | 6–7 years |
| Second molar | 12–13 years |
| Third molar | 17–21 years |
| Lower teeth | Erupt |
| Central incisor | 6–7 years |
| Lateral incisor | 7–8 years |
| Canines | 9–10 years |
| First premolar | 10–12 years |
| Second premolar | 11–12 years |
| First molar | 6–7 years |
| Second molar | 11–13 years |
| Third molar | 17–21 years |

The evolutionary phase does not exhaust the modifications of the swallowing tract because variations continue to occur throughout the entire span of existence, during which some modifications may be transient, such as temporary mouth damage, while others are more stable, such as variations due to iatrogenic injuries. Others also are related to the passage of time: Senescence—the biological process involving functional decline over the years until death—is by no means the age at which variations are most evident and most consequential on the functionality of swallowing (Comfort 1964). As the years progress, changes occur in the ability to swallow of which aging subjects are not always fully aware. These changes are determined by a series of events such as the progressive loss of dental elements, the reduction of muscle tone, the impoverishment of receptor elements, and cognitive decline and salivary deficit (Shaker and Lang 1994; Madhavan et al. 2016; Smithard 2016). In the peripheral nervous system of elderly subjects, there is a reduction in nerve conductivity with a consequent increase in nerve signal conduction times due to the depletion of the myelin sheath (HumbertI and Robbins 2008; Ney et al. 2009). This determines, among other things, a decrease in the efficiency of proprioceptive feedback which is at the origin of the reduction in the ability to discriminate consistencies and viscosity of food (Hiss et al. 2001). Elderly subjects show a reduced sensitivity to all tastes—sweet, salty, bitter, sour, umami (from the Japanese word umai, or "delicious")—and a loss of smell (Vroon et al. 2003; Doty 1981; Engen 1989), which is why current tastes do not correspond to "those of the past," demonstrating not really a modification of foods, or perhaps not only, but an impoverishment of the receptor system. In elderly subjects, an increase in chewing cycles has been observed in relation to the loss of teeth and the presence of dental prostheses. There is also a reduction in salivation as a consequence of anatomical changes in the salivary glands, but more frequently following the intake of pharmacological therapies.

The reported modifications are not necessarily present in all elderly subjects, so there is a significant number of elderly people who maintain good swallowing functionality up to an advanced age (Robbins et al. 1992). The sequence in which the act of swallowing is performed does not change in elderly people, but the times of the different phases are varied. The oropharyngeal phase is prolonged (Butler et al. 2010) with a dilation of the transit times of the bolus (Rofes et al. 2010; Logemann et al. 1998) and with the activation of more cortical areas attesting to a more concentrated effort (Teismann et al. 2010).

The presence of the cough reflex constitutes, as already seen, an important bulwark in defense of the lower airways from the penetration of the bolus. In old age, there is no significant change in the cough reflex, but an increase in the concentration of citric acid required to evoke cough in subjects with dementia is reported ($2.6 \pm 4.0$ mg/mL in control subjects; $37.1 \pm 16.7$ mg/mL in subjects with dementia; >360 mg/mL in subjects who survived aspiration pneumonia; Ebihara et al. 2012). Changes in swallowing function in old age can be determined not only by pathological conditions but also by frailty and sarcopenia. Frailty is "a physiological syndrome characterized by reduced functional reserves and decreased resistance to stress, resulting from the cumulative decline of multiple physiological systems that cause vulnerability and consequent institutionalization and mortality" (Fried et al. 2004). Sarcopenia is "a progressive physiological condition that limits the mass, strength, and function of muscle tissue" (Landi et al. 2018).

The physiological loss of muscle mass begins around 40 years and undergoes an acceleration process around 50–60 years. Both conditions gradually lead to the weakening of skeletal musculature, which results in the reduction of lingual strength and pharyngeal contraction. The reduction of isometric tongue contraction has no consequences as long as it is not in conditions that see it engaged to its maximum potential (Butler et al. 2011). Clavé argues that the reduction of muscle mass of the tongue due to sarcopenia reduces its propulsion force and increases the risk of aspiration (Clavé et al. 2005). Logemann has found a reduction in the strength of the base of the tongue in elderly women (Logemann et al. 2002). Furthermore, the movements of elevation and anterior translation of the larynx are reduced, while the closure of the vestibule and the maximum peak of laryngeal elevation are delayed compared to younger subjects (Butler et al. 2011). Again, in those over 65, there is a reduction in the ability to clean the pharynx, which determines an increase in swallowing acts (Logemann et al. 2002) and the consequent lengthening of the pharyngeal transit time of the bolus. Another cause of changes in the swallowing mechanism at different ages is the consistency of the food consumed. Foods of different consistencies engage the swallowing system differently and imply different modes of activation of the individual anatomical districts. That is, the act of swallowing is consistently composed of four consecutive phases but the level of activation and the ways in which they are carried out differ depending on the consistency of the foods consumed (Table 1.3).

Humans are omnivorous; that is, they are organisms that feed on a wide variety of foods because their digestive system is able to process the different compositions of the diet.

**Table 1.3** Phases of swallowing engaged by type of bolus

| Type of bolus | Oral preparation | Oral transport | Pharyngeal transport | Esophageal transport |
|---|---|---|---|---|
| Liquid bolus | – | ± | + | + |
| Creamy bolus (semi-liquid-semi-solid) | – | + | + | + |
| Solid bolus | + | + | + | + |
| Mixed bolus | Variable | Variable | + | + |
| Secretions | – | – | + | + |
| Refluxes | – | – | + | + |

This characteristic makes it suitable to consume a vast repertoire of different foods that differ in composition and consistency. Such versatility is not always present in extreme ages (infancy and old age), when the swallowing system is not suitable for processing all consistencies. From birth to 4 months, newborns consume a completely liquid diet (milk and/or formula) from the mother's breast or bottle, are unable to consume and digest solid foods (Agostoni et al. 2008), and have reflexes suitable for protecting the swallowing apparatus from solid consistency foods (tongue protrusion reflex, phasic bite reflex, and vomit reflex). At around 4 months of age, the newborn continues to feed mainly on foods of a liquid consistency, but the changes in oral movements that occur in the period 4–6 months, such as the separation of the movements of the tongue from those of the jaw, the desensitization of some protective reflexes, the variation in lingual dynamics, together with the initial ability, albeit with adequate supports, to sit, make swallowing possible and therefore the introduction of foods of a creamy consistency in the diet. From 7 to 9 months, the diet continues to be predominantly liquid, but the newborn is already able to consume a wider variety of solid foods of varying volume. From 9 to 12 months, the child generally consumes a mixed diet that includes both liquid foods and foods of other consistencies (creamy and solid). From 12 to 24 months, the oral management of food becomes more refined until it is progressively possible to process all consistencies effectively. The relationship between swallowing and food consistency is absolutely reciprocal; in fact, anatomical-physiological maturation makes it possible to introduce new consistencies but it is the actual introduction into the diet of new consistencies that develops and refines swallowing abilities. The development of swallowing is therefore the consequence of anatomical modifications, neurological maturation, and exposure to different consistencies. In senescence, the modification of the anatomical conditions makes it necessary to vary the panorama of consistencies used in daily nutrition. The state of conservation of the teeth, but not only, is linked to the possibility of continuing to consume foods with a solid consistency, which are often replaced by foods with a creamy consistency, that is, with a low degree of complexity of oral preparation, as occurs when eating a hamburger rather than a steak.

Another factor that generates variations in swallowing over the course of life is the change in the ways of in which food is taken. The newborn feeds with breast milk or formula by sucking from the mother's breast or bottle. The introduction of creamy foods into the diet also marks the beginning of the use of the spoon for eating. The new way of introducing food into the mouth requires the acquisition of new motor skills, especially by the anterior region of the mouth, for the effective reception of the new instrument that introduces food into the mouth. At this stage, the newborn also experiences contact with the glass, from which he will progressively learn to take liquids as an alternative to the bottle and the mother's breast. The introduction of solid foods, together with the ever-improving ability to maintain a sitting position, promotes the growth of the child's praxic abilities, which progressively allow him to effectively and safely swallow boluses of increasing consistency and volume (Morris and Klein 2000; Delaney 2010; Dodrill 2014). The acquisition of adequate motor patterns continues uninterruptedly and in a certain sense always remains active, to allow the subject to learn all the behaviors necessary to master the various ways of bringing food to the mouth. The adult individual is generally able to eat food with the utensils used by the geographical and cultural group to which he belongs. In the Eastern world, eating with wooden chopsticks is absolutely common, while in the Western world, this method is still a scarcely widespread skill, which began to spread only after the opening of oriental cuisine restaurants. Not only the utensils but also the methods used to welcome food into the mouth are very varied. We can bring food to our mouth with cutlery or with our hands, but also bite it to tear it from a larger piece, or suck it as is sometimes done with spaghetti, or even lick it as is done with ice cream. The repertoire of motor skills that we are able to express for drinking is even more varied. In fact, we can drink from a glass that can be wide or narrow, or from a bottle, but we can also drink directly from a fountain that can have a flow of water that comes down from a tap or gushes from below; we can also drink liquids by sucking them from a bottle (and not just think about newborns because many bottles now have a spout that requires suction to release the liquid they contain) or a straw. When drinking from the bottle while the head is extended, the liquid is poured directly into the pharynx, avoiding any contact with the mouth. All these variables imply the subject's ability to use different motor patterns that, once learned, modify above all the oral preparation phase of swallowing. These abilities vary from one individual to another and are closely linked to cultural and individual variables. The multiplicity of terms with which the wide variety of ways of hydrating oneself is expressed in the Italian language reveals the wealth of oral motor acts that the individual learns during his existence. We list some of them: sipping, slurping, gulping, swallowing, quenching one's thirst, sipping, gulping down, draining, guzzling, tasting, toasting, and sucking.

Swallowing changes throughout life as anatomical, functional, dietary, and cultural conditions vary. Constant phases of swallowing modification can be found in ontogenic development (Table 1.4).

**Table 1.4** Physiological phases of swallowing modification

| Evolutionary phases of swallowing | Swallowing framework | |
|---|---|---|
| Fetal swallowing | Fetophagia | |
| Neonatal swallowing | Pedophagia (0–12 years) | Neonatal (0–6 months) |
| | | Early (7–24 months) |
| Infant swallowing | | Intermediate (2–6 years) |
| Mixed swallowing | | Late (7–12 years) |
| Adult swallowing | Adult swallowing (12–64 years) | |
| Senile swallowing | Presbyphagia (>64 years) | |

## 1.2  Swallowing in Children

The first stages of development occur in the mother's womb, where the swallowing structures form from the brachial or pharyngeal arches (Table 1.5).

Gestation lasts about 40 weeks (range 37–42 weeks). Swallowing movements are one of the first motor manifestations of the pharynx; pharyngeal swallowing movements have been observed between the 10th and 14th week of gestational life (Humphrey 1970). From the 10th week of development of the fetus to birth, the appearance of swallowing activity significantly contributes to the homeostasis of fetal and amniotic fluid. The presence of swallowing during fetal life contributes to the somatic maturation of the digestive and respiratory systems. The functional and anatomical maturation of the digestive system is also evidenced by the increase in esophageal motility and the lower esophageal sphincter, as well as by gastric emptying and the presence of intestinal motility (Kliegman et al. 2011; Gardner and Merestein 2002). From the early stages of development, swallowing shows a wide and articulated repertoire of movements, whose variety and multiplicity have been highlighted and classified by ultrasound studies. Miller et al. (2003), based on ultrasound observations, have identified the following fetal swallowing movements:

- *Tongue cupping*: Raising of the lateral margins of the tongue, which generally precedes a swallow.
- Repetitive *licking* movements: Of fingers, uterine wall, placenta, umbilical cord, etc.
- *Suckling*: Rhythmic movements of the tongue in an anteroposterior direction, following contact with other structures, with ingestion of amniotic fluid in the oral cavity.
- *Mouthing*: Rhythmic opening and closing movements of the mouth.
- *Crunching*: Vertical movements of the jaw with flow of amniotic fluid.
- *Swallowing*: Of a bolus of amniotic fluid with lingual propulsion into the pharynx and pharyngeal contraction.
- *Glottic flutter*: Rapid adductions of the arytenoid cartilages.
- *Hiccups*: Rhythmic and rapid movement of fluid accompanied by diaphragmatic contraction.

**Table 1.5**  Fetal development from 8 to 40 weeks

| 0–8th week | Formation of oropharynx, esophagus, trachea, and taste buds begins |
| --- | --- |
| 10th–11th week | First swallowing acts |
| 12th–13th week | Functionality of taste buds |
| 18th–24th week | The fetus begins to suck (swallowing about 2 liters of amniotic fluid per day) |
| 26th–29th week | Appearance of oral reflex movements |
| 32nd–34th week | Structuring of the sucking-swallowing coordination |
| 34th–35th week | Sufficient swallowing ability for breastfeeding |

- *Inhalation*: Rhythmic and slow passage of fluid through the pharynx into the trachea.
- *Oral breathing*: Isolated jet of amniotic fluid that propagates through the oral cavity, in and out of the cavity, of amniotic fluid at rates of 3–5 s.
- *Nasal breathing*: Isolated jet of amniotic fluid that propagates through the nose, in and out of the cavity, at rates of 3–5 s.
- *Coughing*: Rapid, single, and isolated movements of nostrils-mouth accompanied by a rapid forward movement of the head.
- *Snoring*: A short, rapid, isolated jet of fluid from the nose.

Swallowing varies according to changes in the nervous system, following its maturation. The central nervous system matures in a bottom-up sequence (Kenner and McGrath 2010; Stiles and Jernigan 2010): In the first trimester of fetal life, the first synapses form at the spinal level. In the second trimester, the brainstem matures. The primordial respiratory reflexes (rhythmic contractions of the diaphragm and thoracic muscles) and the first sucking and swallowing begin to emerge in parallel with the maturation of the brainstem structures. In the third trimester, however, the volume and surface of the brain increase markedly. In this period, the first learnings are possible, as evidenced by the fact that newborns are able to recognize sounds and smells to which they were exposed during fetal life. With birth, life begins in an extrauterine environment, where the newborn must deploy all its resources in order to undertake an autonomous life, although it still has the utmost need for parental care, since the maturation of the CNS is largely incomplete and the cerebral cortex in particular is still immature. At this age, the brainstem is the most developed area of the CNS and is able to exhibit a series of respiratory and swallowing reflexes essential for survival. The cerebral cortex follows a maturation path that develops over a period of about 5 years. The neuronal equipment, already largely formed in the prenatal period, is poorly connected at birth. The cerebral cortex, unlike the trunk and spinal cord, produces the largest number of connections in the postnatal period (Stiles and Jernigan 2010). At 2 years of age, a child's brain contains more than 100 trillion synapses that develop at a differentiated growth rate in the various cortical areas. Synapses develop primarily in sensory areas and ultimately in areas involved in cognitive and emotional functions (frontal and temporal lobes). The number of new synapses remains high in all cortical areas until the age of eight, when it tends to decrease and settle at a growth rate equal to that of adults.

From a strictly deglutological point of view, the period commonly called "pedophagia" begins at birth. Pedophagia defines swallowing between the ages of 0 and 12 years, when adult-type swallowing begins. This phase is commonly divided into four stages that develop at different ages and are therefore characterized by different swallowing modes (Table 1.6).

Swallowing is commonly divided into three phases: oral, pharyngeal, and esophageal. The oral phase is the one that undergoes the most marked variations during the evolution of swallowing. In fact, if in adults and children the oral phase is a voluntary one, in newborns it is involuntary so that at this age, all the phases of swallowing are involuntary (Morris and Klein 2000; Wolf and Glass 1992). Swallowing at birth is conditioned by the anatomical characteristics of the swallowing tract, the type of food, the ways in which the boluses are taken, and the characteristics of the CNS. As already partly seen at birth, the cerebral cortex is largely immature, while the brainstem presents a series of reflexes that are used in the realization of the act of swallowing. The oral reflexes present at birth can be divided into adaptive and protective. Adaptive reflexes allow the newborn to direct the bolus toward the intestine, while the protective reflexes keep food away from the airways. Reflexes are activated by adequate stimulation and are dependent on the alertness and hunger of the child (Morris and Klein 2000; Dodrill 2014; Table 1.7). Over time, reflexes tend to be replaced by more sophisticated voluntary behaviors, with the exception of the cough and vomiting reflexes, which continue to exert their protective function throughout life.

**Table 1.6**  Pedophagia: subdivision into stages

| Neonatal pedophagia | 0–6 months |
|---|---|
| Early pedophagia | 7–24 months |
| Intermediate pedophagia | 2–6 years |
| Late pedophagia | 7–12 years |

**Table 1.7**  Oral reflexes

| Adaptive | Protective |
|---|---|
| Suckling reflex (extinguishes at 3–6 months) | Tongue protrusion reflex (extinguishes at 3–6 months) |
| Cardinal points reflex (fades at 3–6 months) | Tongue rotation reflex (extinguishes at 6–9 months) |
| | Phasic bite reflex (extinguishes at 9–12 months) |
| | Cough reflex (does not extinguish but continues to be present throughout life) |
| | Vomiting reflex (does not fade but between 6 and 9 months, the areas of elicitability change and remain highly individual based, probably, on subjective sensory experiences) |

Source: Morris and Klein (2000), Dodrill (2014), and Logemann et al. (1998)

In the newborn, the predominant role is undoubtedly assumed by the sucking reflex (*suckling*), which is elicited by stimulation of the tip of the tongue and the anterior part of the palate. In response to stimulation in these areas, the tongue moves back and forth on a horizontal plane. The sucking reflex appears early in the third trimester of gestational life and remains active until 3–6 months of age of the newborn (Morris and Klein 2000; Dodrill 2014), when it is replaced by a more mature suckling pattern of a voluntary nature, sucking.

The evolution from suckling to sucking is not only the transition from one sucking mode to another, made possible by the changed anatomical conditions of the oral region, but is above all the testimony of the development of the CNS. At about 6 months, the maturation of the newborn's cerebral cortex makes possible the voluntary control of sucking which is no longer therefore under the exclusive control of the brainstem and therefore, from a completely reflex act (*suckling*), becomes a voluntary act (*sucking*). The maturation of the cerebral cortex makes the newborn capable of making decisions and controlling his own motor patterns. From an anatomical point of view, the transition from one form of sucking to another is made possible by the enlargement of the dimensions of the oral cavity, which allows wider movements of the tongue, which in this phase begins to move independently of the jaw. In fact, if in the suckling phase the movements of the tongue are contextual with the jaw and develop on a horizontal plane exclusively along the forward-backward direction, in the sucking phase, the tongue moves independently from the jaw and develops more varied and wide movements also on the vertical plane (Dodrill 2021). Toward 6 months, the transition from suckling to sucking determines the possibility of swallowing boluses of a nonexclusively liquid consistency. This change in swallowing is accompanied by the child's ability to maintain a sitting position, the beginning of tooth eruption, the progressive disappearance of neonatal reflexes, the progressive development of fine motor skills, and finally, the beginning of the learning progress toward independent walking. In this phase, in addition to the sucking that serves to provide the newborn with the nutritional elements necessary for survival and growth (nutritive suckling), there is also a nonnutritive suction (nonnutritive suckling) which has the purpose of reassuring the newborn in critical moments. Nutritive sucking is used during feeding and the ratio between sucking and swallowing is about 1:1, while in nonnutritive sucking, used for calming purposes, the ratio between sucking and swallowing is 6:1 or 8:1.

The next phase of pedophagia is the intermediate one (2–6 years) in which we see the improvement of the oral phase which, as a consequence of anatomical changes (widening of the internal diameters of the mouth, eruption of the deciduous teeth), expresses increasingly effective chewing behaviors, characterized by fewer and faster chewing cycles (Table 1.8).

In the intermediate phase, we see the coexistence of adult swallowing modalities with infantile ones. In fact, the tongue, which now has a more adequate space available to express its movements, begins to transport the boluses toward the pharynx using the up-down dynamic more and more often, which progressively replaces the in-out dynamic, typical of infantile swallowing. In this phase, the anticipatory phase

**Table 1.8**  Time (seconds)

|                | 2 years | 3 years | 4 years | 5 years | 6 years | 7 years | 8 years |
|----------------|---------|---------|---------|---------|---------|---------|---------|
| *Reduction in duration* | | | | | | | |
| Cracker        | 18.4    | 14.2    | 13.8    | 14.4    | 12.8    | 13.3    | 12.6    |
| Apple juice    | 6.3     | 5.4     | 4.4     | 3.4     | 2.2     | 2.3     | 2.3     |
| *Reduction of masticatory cycles* | | | | | | | |
| Cracker        | 19.7    | 17.1    | 17.6    | 16.7    | 15.8    | 16.2    | 15.9    |
| Apple juice    | 4.5     | 3.9     | 3.5     | 3.0     | 1.8     | 1.8     | 1.7     |

stabilizes and the child shows his tastes in food more effectively and the first food phobias begin to emerge. Children often prefer foods prepared in specific ways and may begin to show neophobias, real phobias toward new foods (Dodrill 2021). During the late pedophagia phase (7–12 years), lingual thrust disappears permanently, and by 10 years of age, 70% of children have achieved adult-type swallowing. Chewing reaches a higher degree of efficiency, thanks to the greater completeness of the dentition and the further expansion of the oral cavity.

The tongue at rest maintains itself on the retroincisal papilla, progressively abandoning the position on the oral floor. The larynx also completes its descent into the neck, while taste and smell further mature.

## 1.3   Swallowing in Adults

Swallowing in adults follows a model considered universally valid although it is important to take into account the great interindividual variability found among normal subjects, especially in the preparation and oral transport phases (Stephen et al. 2005; Hiiemae and Palmer 1999). The variability is linked to different factors, such as the type of food, the context in which the subject eats, and last but not least, the swallowing assessment tool used. Swallowing ensures the safe transport of food to the stomach in such a way as to prevent the accidental passage of the bolus, or parts of it, into the lower airways. To achieve this goal, swallowing must perform two tasks: make the bolus suitable for transit inside the swallowing canal and then move it quickly and safely from the lips to the stomach. Oral preparation serves to adapt the characteristics of food to those of the swallowing canal, within which the bolus will have to transit without damaging it. The sliding of the bolus inside the swallowing canal is produced by the differences in pressure gradient created by the muscles of the swallowing canal upstream and downstream of the bolus itself. In other words, upstream of the bolus, a high-pressure zone is created by the contraction of the musculature of the anatomical tract involved, while the relaxation of the musculature located downstream of the bolus determines a low-pressure zone that favors its progression. When due to pathological conditions (neuromuscular deficit or incoordination of the different muscular districts) this pressure gradient difference is not created, the bolus remains blocked, because it is unable to continue its journey to the stomach, in the usually short times (Groher 2021).

Adult swallowing is also divided into three phases: oral, pharyngeal, and esophageal. The oral phase is further divided into the oral preparation phase and the oral transport phase (Table 1.9).

The oral preparation phase takes place entirely within the oral cavity, where the bolus is made suitable for transit in the subsequent portions of the swallowing canal (Schindler 1990) and especially "felt" in its sensorial qualities (Groher 2021). The taste qualities of foods are appreciated especially in this phase (Schindler et al. 2011), in which saliva plays an important role in the perception of flavors, activating the taste receptors located on the tongue, on the hard palate, on the soft palate, and in the pharynx (Groher 2021). In the oral preparation phase, the motor patterns of the mouth change depending on the consistency of the bolus (Schindler et al. 2011), while in the swallowing of solids and liquids, there are no significant differences in the pharyngeal and esophageal phase, even if the greater consistency of solid boluses determines the increase in the time it takes to pass through the pharynx and esophagus and the possible permanence of residues (Matsuo and Palmer 2013). Therefore, the most significant differences between swallowing solids and liquids occur in the oral phase, in which chewing, in a highly coordinated manner with oral transport movements, precedes the movement of solid boluses toward the pharynx. Solid boluses need to be broken up and crushed by the teeth and by the combined action of the jaw and the tongue which, thanks to its rich repertoire of movements, is able to move the bolus from one dental arch to the other. The tongue also participates in the deformation of the bolus with the pressure it is able to exert against the hard palate. Traditionally, it has been believed that the bolus remained in the mouth during the oral preparation phase, while the pharyngeal phase was activated when the bolus arrived at the isthmus of the fauces (Logemann et al. 1998). A model born from mammalian feeding was subsequently extended to humans (German et al. 1989; Hiiemae 2000; Herschel et al. 1984; Franks et al. 1985; Thexton and Hiiemae 1997; Hylander et al. 1987), or the so-called "feeding processing model" (Palmer et al. 1992), which illustrates chewing and the oral phase of swallowing. This model describes different oral processing modes depending on the consistency of the bolus. Solid boluses are processed orally in three different phases: first transport phase (pull-back), chewing phase (food processing), and second transport phase (squeeze back). When the bolus enters the oral cavity, it is positioned on the upper face of the tongue, which transports it to the premolars and molars and then, by rotating on itself, positions it between the occlusal surfaces of the two dental arches

**Table 1.9** Swallowing phases

| Phases | | Actions |
|---|---|---|
| Oral phase of: | Preparation | Mastication |
| | Transport | Movement of the bolus from the front of the mouth to the valleculae |
| Pharyngeal phase | | Movement of the bolus from the valleculae to the upper esophageal sphincter (UES) |
| Esophageal phase | | Movement of the bolus from the SES to the lower esophageal sphincter (LES) |

(oral transport phase 1 or pull-back). The chewing phase (food processing) immediately follows transport phase 1 and reduces the bolus into smaller particles, lubricates it, and reassembles it with the aid of saliva. In this phase, the mandible is in continuous movement, stopping briefly only during the swallowing phase. The movements of the mandible are closely connected with the movements of the tongue, the soft palate, and the hyoid bone. The soft palate, during the chewing phase, moves cyclically in opposition to the movements of the mandible; that is, it rises when the jaw opens and vice versa it lowers, thus opening the access to the nasal cavities, when the mandible rises (Matsuo et al. 2005; Matsuo et al. 2010). This dynamic determines the absence, during chewing, of a stable separation between the oral and pharyngeal cavities (Hiiemae et al. 1996; Matsuo et al. 2005) and favors the retronasal sense of smell. The raising of the mandible during the chewing phase determines the reduction of the volume of the mouth and the consequent thrust of the air contained in the mouth toward the nasal cavity through the retronasal route, favoring the stimulation of the chemoreceptors located in the rhinopharynx and in the nasal cavity, which allow the perception of the aroma (Hodgson et al. 2003; Buettner et al. 2001). During chewing, the large lingual movements in the vertical and anteroposterior dimension are closely coordinated with the movements of the mandible. The forward movements of the tongue coincide with the opening movements of the jaw. Vice versa, the posteriorization movements of the tongue coincide with the closing movements of the mandible. During chewing, the tongue does not perform movements exclusively on the anteroposterior axis but also on the vertical and mediolateral ones to maintain the boluses under the occlusal face of the teeth and promote their comminution. To maintain the bolus under the occlusal surface of the teeth, the cheeks also work together, which in turn push the pieces of food that have ended up in the gingival sulci back into them.

When a part of the chewed bolus is ready to be swallowed, it is moved to the surface of the tongue where, through a squeezing mechanism created by the surface of the tongue in opposition to the hard palate, it is pushed toward the oropharynx (second phase of transport). During the second phase of transport, the tongue moves upward while the mandible moves downward, and vice versa, the soft palate moves upward to close the access to the nasal cavities (Hiiemae and Palmer 1999; Matsuo and Palmer 2008).

Phase 2 of transport occurs intermittently during chewing (Matsuo et al. 2005), and the transported food accumulates on the oropharyngeal face of the tongue and in the valleculae until it reaches, through the subsequent phase 2 of transport, a critical quantity that activates the swallowing reflex. The bolus accumulation times see a large interindividual variability. In fact, in some subjects, it lasts a fraction of a second, and in others up to 10 s (Hiiemae and Palmer 1999).

Liquids do not require a chewing phase and therefore, in this case, transport phase 2, which occurs in a similar way to solids, immediately follows transport phase 1 (Table 1.10).

The pharyngeal phase of swallowing is undoubtedly the most delicate, because it is at this level that the airways and digestive tracts intersect, creating a dangerous intersection that various defense mechanisms make safe. The bolus, positioned in

**Table 1.10** Swallowing liquids and solids

| *Swallowing liquids* | | | |
| --- | --- | --- | --- |
| Transport phase 1 | Transport phase 2 | Pharyngeal phase | Esophageal phase |
| *Swallowing solids* | | | | |
| Transport phase 1 | Mastication | Transport phase 2 | Pharyngeal phase | Esophageal phase |

the back of the mouth, is pushed toward the pharynx by the movement of the tongue raising toward the hard palate and the soft palate. When the bolus arrives, the lateral walls of the pharynx medialize, thus forming a funnel that guides the bolus toward the hypopharynx. At the same time, the soft palate moves upward to close access to the nasal cavities, aided by the lateral and posterior walls of the pharynx. In this phase, there is a high risk that the bolus, or parts of it, may enter the lower airways, whose protection is therefore ensured by three different mechanisms: the forward and upward displacement of the hyoid bone and the larynx, the closure of the glottis, and the overturning of the epiglottis (Martin-Harris et al. 2003; Hårdem Ark Cedborg et al. 2010). The coordination between breathing and swallowing is in fact a further system for protecting the lower airways. In fact, swallowing acts, although interindividual differences have been detected, occur during the exhalation phase, which is interrupted during swallowing to resume immediately after the passage of the bolus (expiratory-expiratory pattern) (Andrew 1956; Paydarfar et al. 1986; Martin-Harris 2000; Doty 1968; Martin et al. 1994a; Perlman et al. 2005). Moreover, the system is further protected by coughing, an important reflex for the removal of secretions or accidentally inhaled materials from the bronchial tree. In the pharynx, four different mechanisms regulate the progression of the bolus toward the esophagus: the lingual push, the contraction of the pharyngeal walls, the negative pressure of the hypopharynx, and gravity (McConnel et al. 1988; Dodds 1989; Dodds et al. 1990). The opening of the upper esophageal sphincter marks the end of the pharyngeal phase of swallowing.

The third phase of swallowing is the esophageal phase which, last in chronological order, effectively concludes the act of swallowing by initiating the digestive phase.

The esophageal phase begins with the opening of the upper esophageal sphincter (upper esophageal sphincter, SES) which is composed of the lower pharyngeal constrictor muscles, the cricopharyngeus muscle, and the most proximal part of the esophagus. The UES is closed at rest by tonic muscle contraction and opens with the contribution of three factors: the relaxation of the cricopharyngeus muscle, which precedes the opening of the UES or the arrival of the bolus; the contraction of the suprahyoid and thyrohyoid muscles that move the hyo-laryngeal complex upward and forward; and the pressure of the descending bolus (Cook et al. 1989; Ertekin and Aydogdu 2002). Once they enter the esophagus, solid and liquid boluses are conducted into the stomach by sequential (peristaltic) contractions of the circular muscles of the organ, by the appropriate opening of the sphincters that, at rest, close the esophagus at the top and bottom, and by the shortening of the esophagus following the contraction of the longitudinal muscles. The contraction of the esophagus is primary peristalsis: When the bolus enters the esophagus, the upper esophageal sphincter closes and a peristaltic wave is activated that reaches the already relaxed lower esophageal sphincter, which contracts when the peristaltic wave arrives. The peristaltic wave is made of two main parts: an initial relaxation wave that welcomes the bolus, followed by a wave of contraction that pushes it toward the lower esophageal sphincter lower esophagus, which, closed at rest to prevent regurgitation from the stomach, releases during swallowing to allow the passage of the bolus into the stomach. The movement of the bolus toward the stomach is further aided by the gravitational attraction that is exerted on the bolus itself when the subject eats standing or sitting.

## 1.4  Swallowing in Older Persons

Advancing age brings about changes in swallowing as well as in other body districts. Senile swallowing represents the adaptation of swallowing functionality to the aging of the organs involved in swallowing, in relation to the anatomical and functional changes that occur with the progression of age. However, such changes do not necessarily imply a worsening of swallowing, which often remains functional throughout life (Robbins et al. 1992; Ship 1999). Functional changes affect both sensory and motor aspects (Robbins et al. 1992, 1995; Feldman et al. 1980; Cook et al. 1994; Shaker and Lang 1994; Logemann et al. 2000; Nicosia et al. 2000; Daniels et al. 2004). Taste and smell tend to decrease with age, although in a differentiated manner; many studies highlight how in the elderly the loss of smell proceeds more rapidly than the loss of taste (Cain and Stevens 1989; Stevens and Cain 1985). Somatic sensitivity, on the other hand, is relatively preserved in the elderly (Fukunaga et al. 2005; Calhoun et al. 1992; Petrosino et al. 1982). Chewing is affected by the effects of time mainly in relation to the tooth loss; in fact, in elderly people who retain their teeth, there are no chewing difficulties (Kossioni and Dontas 2007; Heath 1982; Carlsson 1984; Helkimo et al. 1978). Saliva plays an important role in digestion, oral health, and swallowing, which is why its presence in the elderly plays an important role in preserving swallowing function. Some studies report a substantial stability in saliva production in old age (Shern et al. 1993), while others report a reduction in salivary flow (Percival et al. 1994). It is generally believed that saliva production in the elderly is sufficient for the normal performance of oral functions, although approximately 25% of the elderly suffer from xerostomia (Osterberg and Carlsson 1979) following Sjögren's syndrome, radiotherapy in the head-neck district, and in relation to the intake of drugs, since the amount of salivary flow is inversely proportional to drug intake (Narhi 1994; Atkinson et al. 1990; Liu et al. 1990). The musculature is affected by the progressive loss of muscle fibers (sarcopenia) and the orofacial musculature is not spared from this process. This process does not spare the orofacial musculature. The isometric lingual pressure, which is so important in chewing and in pushing the bolus toward the pharynx, is reduced in its maximum values, although it remains able to produce pressure values adequate for the progression of the bolus. A study that compared the swallowing of young and elderly subjects found a reduction in the maximum values of isometric tongue pressure at three different points of the tongue and a greater latency to reach the peak of pressure but a substantial identity of values during the swallowing of the bolus (Nicosia et al. 2000). This demonstrates the reduction of the functional reserve, which exposes the elderly to dysfunction in case of exceptional events. The longer latency in reaching the maximum pressure peaks could explain the difficulty that elderly people have in swallowing liquid boluses which, due to their high rheological properties, pass through the various areas of the tongue before they reach the pressure peaks necessary for the correct and safe performance of the swallowing act.

# Executive Attentive Networks, Motor Learning, Dual Tasks

2

The ABAED® method is developed in various phases that require in-depth knowledge of some steps connected with the neuroscientific theories that support them. It must be emphasized that "enabling" means making someone learn what has not been achieved with typical development. However, any cognitive or motor learning is modulated through interaction with different processes that involve emotional motivational systems, attention, the so-called "executive functions" and working memory. Therefore, basic knowledge of these systems (and especially their interaction) is important if not fundamental for those who want to act, enable, or train with awareness and competence in any context, whether clinical or in school or sports life. In this chapter, some minimum necessary foundations will be illustrated, still useful for trainers. We will start with the basic notions of attentive networks and working memory with the possible interactions with motivational and emotional systems. The executive attention bases of motor learning will follow and finally, we will mention the variety of uses of dual tasks. The goal is to guide the trainer toward the salient points useful for achieving effectiveness in the application of the method. However, a certain relevance of results can be evaluated that the method, downloaded from the type of pathology (i.e., not necessarily related to swallowing but considered more generally in disorders and in very different applications), has already provided. In fact, strong signals of incisiveness in enabling and rehabilitative improvements have been detected (Benso 2004a, b; Benso et al. 2008a, b; Ciarmiello et al. 2015; Veneroso et al. 2016; Benso et al. 2021).

## 2.1 Attentive Networks

The term "attention" is very common, but it takes on different meanings in the situations and in the intentions of those who use it, also based on the knowledge and competence that each person may have. There are many types of attention and experts in the field also know how to isolate them (and point them out) in everyday



A. Amitrano, F. Benso, *Atypical Deglutition Treatment*,
https://doi.org/10.1007/978-3-032-01482-5_2

life as well as treat them experimentally in the laboratory. We have moved from a historical phase in which the existence of attention in certain processes was even denied (in which it has now been discovered that it is fundamental) to a phase in which there is an excess of citations of unclear attentive processes and uncertain executive functions. To know how to learn (and therefore enable), we need to know the neural models underlying the different learning phases, both cognitive and motor. An important neuronal substrate in this sense is provided by the complicated interrelations that exist between different attentive networks and motivational emotional systems. Trying to isolate models that can minimally represent this collection of networks (typical of brain complexity) is an effort that neuroscience is tackling with initial very interesting results, especially in perspective.

In addition to the complexity, there are several obstacles that arise in dealing with this knowledge, such as the repeated use of largely outdated models that continue to rage in hundreds of publications. This is the practical manifestation of the inertia described by the epistemologist Thomas Kuhn (1922–1996) in receiving a paradigm shift after numerous falsifications of a theory (Kuhn 1985). As a result, reduction and simplification operations are repeated in the attempt to make linear, determined, and established what is not by nature.

Going in order, we will briefly see some of the most important attentive networks and then to better assimilate the concepts, we will use a story that describes the theories of executive attention through life episodes. In this sense, we will understand how, by deepening these theories, real events can be recreated and read according to an executive attentive "score." This ability to transfer and analyze neuroscientific models even in routine life situations is what, subsequently, allows the expert to isolate clues in problematic situations and to interpret particular facts useful for unraveling apparently very complicated clinical or existential situations. In other words, this knowledge becomes essentially useful to support the application activities of coaches, teachers, and clinicians.

As suggested by the most recent models, the functional study will proceed by evaluating the circuits, including the most influential areas (defined as "hubs") inserted in the networks themselves. These hubs seem to provide a decisive contribution for the performance of particular tasks. However, the neuroscientific reality does not stop at this level and, in these cases, it leads to evaluate two other essential points. First, the single areas, even if definable as "hubs," cannot operate fully without the contribution of the other components of the network considered; therefore, the entire "circuit" now remains the functional protagonist. Second, the final result, after having performed certain cognitive or motor tasks, will also depend on a complex interaction between executive attentive circuits, motivational and emotional, always involved. The variability further increases by considering the activation profiles that can change from task to task, from subject to subject, and even from mental state to mental state of the same subject. This second point makes us reflect on how misleading it is to think of isolating specific executive functions with tasks that inevitably involve several others (known and unknown) as well as working memory, without neglecting the influence of emotional circuits and servosystems that include input analyzers and output processors.

### 2.1.1  The Default Mode Network (DMN)

There are moments when we are not particularly busy mentally, thoughts run freely, and in this state, the variability of sensations (negative or positive) is the rule. Considering the positive moments, we can say that they reflect experiences in which we could enjoy intuitions or pleasant internal sensations, dictated by autobiographical memories. Even various external signals received in this particular state of "free mind" can have a greater impact than in other moments. It is the "magical" condition when we experience sublime atmospheres while listening to music, in the presence of grandiose works of art, or in front of natural or human phenomena that are satisfying for us. Considering, instead, the negative moments, this state transforms into depressive tendencies anxiously accompanied by thoughts of worry or, even worse, by "rumination of thought" often characterized by particular obsessions that appear repeatedly. The state of "free mind," if not educated, tends more frequently toward negative, "destructive" emotions, as the Dalai Lama and Goleman (2009) would say. Neuroscience has framed the circuits of these networks that, indeed, can have positive or negative value and that must therefore be known and monitored also as the cause of various successes or failures that then again fall into the experiences made in the clinic or in everyday life.

In neuroscience, these circuits are called as Default Mode Network (DMN). DMNs are a set of brain areas that are activated when subjects are not engaged in cognitively complex tasks, as we have just outlined. This brain activity has been evaluated as a default activation, and at the beginning of their identification, the DMNs were characterized as a state of wandering thought. The definition of DMN was introduced for the first time by Raichle et al. (2001), when they observed that a set of neural regions showed greater activity when an individual was placed in a sort of cognitive rest. In particular, these networks include the precuneus, the posterior cingulate cortex, the mesial prefrontal cortex, the lateral parietal cortex, and the mesial temporal lobes (Greicius et al. 2003; Fox et al. 2005; Uddin et al. 2009; Greicius et al. 2009). In recent years, there has been an increase in scientific interest for the DMNs, both in typical and clinical populations. These brain regions have been shown to be deactivated during a variety of neuroimaging activities that require cognitive processing, especially in conditions that required particular attention directed to the task. However, in subjects who show some disabilities and distractibility, the DMNs are activated even when cognitive attention tasks require their deactivation (Fassbender et al. 2009). In this case, they negatively influence the focus on the task (intrusive, depressive thoughts, which create a background disturbance in concentration on the performance).[1]

---

[1] Over time, insights have emerged that have isolated more branches of the initial circuit. The DMNs would be anatomically divided into two main branches that converge in an important integrated node mainly comprising the posterior cingulate cortex. In the first branch, the medial temporal lobe subsystem provides information from previous experiences in the form of memories, even autobiographical ones, and associations that are the building blocks of mental simulation. In the second branch, the medial prefrontal subsystem facilitates the flexible use of this information during the construction of self-relevant mental simulations (Andrews-Hanna 2012).

## 2.1.2   The Central Executive Network (CEN)

Let's now move on to a second network, which has very different functions from those described for the DMNs and has been known for many years, during which it has changed its appearance. It deals with the control functions that were known as "frontal functions," indicating the abilities lacking in case of lesion of the frontal lobes (Luria 1977). This purely neuroanatomical terminology was later replaced by definitions of a more "mental" level. Two factors have mainly influenced this: Not all frontal functions are attentional executive; the brain areas that support the so-called executive functions extend beyond the frontal lobes, including in addition to these above all the parietal lobes, the basal ganglia, the insula, and the cerebellum. In 1973, Pribram coined the definition of "executive functions" (EF) and since then, this expression has gradually, but increasingly, entered the phraseology of many authors: Currently, it is in fact a jargon expression, not always precise, used to approximately identify what were once called "frontal functions." The executive attentive network we will discuss is the one that includes and supports the different executive functions and working memory and has been called by several authors Central Executive Network (CEN).

The CENs are activated during tasks that require control, self-regulation, and concentration and are anatomically represented by the dorsolateral frontal and parietal cortex with subcortical couplings (basal ganglia and cerebellum). The CENs are engaged in higher-order cognitive and attentive control (Menon and Uddin 2010; Petersen and Posner 2012; Tang et al. 2012). This system integrates the Executive Attention (EA; Engle and Kane 2002; 2004) system mainly represented by the dorsolateral prefrontal cortex (DLPFC) in which the Working Memory Capacity (WMC), one of the most complete models of working memory, is expressed.

The integrity and evolution of the CENs is fundamental in order to develop attitudes, learning, and self-regulatory aspects; moreover, an efficient working memory is useful and indispensable for all complex activities (cognitive or motor) that require concentration of attentive resources to be learned.

## 2.1.3   The Salience Network (SN)

A third network has recently been introduced and is beginning to shed light on various aspects and to answer questions that have remained unanswered for decades: Who promotes changes in attentional state? Who "accumulates evidence" to direct a certain behavior based on the salience of external or internal stimuli? Some answers to these questions are beginning to be identified. In any case, it is a real revolution with respect to knowledge on executive attentive functions, still to be fully discovered, which brings out new perspectives of framing for different types of pathologies, such as autism, schizophrenia, depression, attentional weakness, and frontotemporal dementia (Uddin 2014; Menon 2015). In this case, further explanations of the possible weaknesses found emerge that require new interpretation. These networks, called salience networks or Salience Network (SN) would have the peculiarity of promoting changes in state between DMN and CEN based on the signals received from internal and external systems.

The third circuit involved in network interactions is therefore represented by the Salience Network (Dosenbach et al. 2008; Sridharan et al. 2008): It is formed by the anterior insula (IA) and the anterior cingulate cortex and has extensive connectivity with subcortical and limbic structures involved in reward and motivation. It is also important for monitoring the salience of external inputs and internal brain events (Sridharan et al. 2008; Menon 2015). The SN, and the IA in particular, plays a critical and causal role in the passage between the CEN and the DMN, through activity paradigms and modes that suggest a causal and potentially critical role for the IA in cognitive control (Dosenbach et al. 2008; Menon and Uddin 2010). In other words, the switch from DMN to CEN is triggered by the SN and this implies the ability to maintain sustained attention on the task to be performed and limits the inevitable falls in DMN. The physiological correlation between these two networks is negative: When one between CEN and DMN is activated, the other is deactivated and vice versa (Kelly et al. 2008); this correlation could be interrupted in synchronization, probably promoted by the SN, in subjects with attention difficulties. However, the interaction between these networks is still little known along with the frequency of the change of activation, which shows substantial variability over time and context (Dixon et al. 2017).

## 2.2 Orientation and Alerting Networks

The complexity, when dealing with "attention," is given by the fact that it is not a linear system made of a few defined components. Even leaving aside various other aspects, we should still introduce alertness and orientation, in order to explain a useful intervention model for treatment, including that of swallowing. We will do this briefly, and to better clarify the often misunderstood concepts of "executive attention" and "working memory," we will include inserts for clarification and in-depth analysis. Finally, we will try to present a contextualizing summary, through a fictional story in which the different attentive nuances will be highlighted. The aim is to make what is substantial to support intervention models in various cognitive/motor realities understood. It is also hoped that the complexity of the networks, thus described, can be better understood by those who approach the subject for the first time or are only stuffed with theoretical studies, never applied in the laboratory or in the clinic. Let's now deal, very briefly, with the orientation and alert networks.

### 2.2.1 The Orienting Network (ON)

The Orienting Network (ON) can support shifts of visual, auditory, and tactile attention in a voluntary or automatic way. The voluntary orientation of attention is supported by the activation of the frontoparietal cortex and the intraparietal sulcus (Petersen and Posner 2012): In particular, the frontal eye fields, useful for the voluntary movement of the eyes, contain neurons dedicated to the voluntary orientation of attention, which is a top-down process.

Instead, the circuit that includes the locus coeruleus, the temporoparietal junction, and the ventral frontal cortex is involved in automatic orientation (Corbetta and Shulman 2002). There are studies that confirm and disconfirm the fact that the latter would be the network characterized by noradrenergic neurons that, ascending from the locus coeruleus, branch out toward the right (temporoparietal and ventral frontal junction). From these observations also arises the hypothesis of cases of hemiinattention or neglect for the space contralateral to the lesion that is predominantly on the right and would affect the noradrenergic circuit (ibid.).

Automatic orientation is innate and, therefore, is guided by stimuli (bottom-up process) and prevails in the first months of life, then remains active even in adults. The voluntary system develops over time also through environmental stimuli. The automatic component can always interrupt the voluntary one in progress, while the opposite is not true. Interrupting the voluntary application (even very concentrated) with the appearance of sudden stimuli in the receptive fields can be of vital importance if such stimuli that capture attention are constituted by dangers or vital survival clues.

## 2.2.2  The Alerting Network (AN)

The Alerting Network (AN) in literature often overlaps with automatic orientation, hence the uncertainties of various studies, as reported by Petersen and Posner (2012). The alert is divided into "phasic" (short time between the alert and the target) and "tonic" (longer time between the alert and the target). The preparation time to the response that starts with the warning is called warning and is represented by a "U" curve which can represent slower response times when one is at the beginning or at the end of the waiting interval.

The network would be characterized by the activation of the right hemisphere, linked to sustained vigilance, as expressed above for orientation, starting from the locus coeruleus, then to the temporoparietal junction, and then to the ventral frontal cortex (Corbetta and Shulman 2002). Petersen and Posner (2012) highlight different points of view that are not in agreement and not easily judgeable based on what is known so far.

As far as we are concerned, the phenomenon of phasic alertness becomes interesting, which puts subjects in a particular state of attentive waiting and constitutes a point of certain interest for treatment, just like when a "ready" precedes a "go," you can think of a sprinter on the starting blocks who hears the signal and activates, with his U-shaped warning (preparation) curve. If the "go" comes too early, the athlete tends to remain still on the blocks; if instead it comes too late, it can promote false starts (anticipations to the target, detected by those who work with reaction times): For us, phasic alertness was an important point of study, shortly before and shortly after the year 2000, for the application that we would later make of it in the clinic.

We were saying that phasic alertness, an innate process (Tang et al. 2012) proposed with various exercises (motor or cognitive), can be thought of as a sort of "ready, go." These exercises can be used to activate the attentive systems that are disarmed by stress and demotivation, or to compensate for the weaknesses of sustained attention, which must be developed over time. In this case, by not requiring the weak subject to "pay attention" but by working on his alert system and adding many "ready, go," the subject can be trained to a more prolonged tonic alertness, without explicitly asking him to "pay attention" (a request that in these types of subjects is made too often during the day, without ever being able to be fully satisfied). Therefore, already around the year 2000, we were requiring the interaction between the focus of attention, the warning curves (Benso et al. 1998), and the different warning styles, changing sensory modality during the attentional shifts between audition and vision (Turatto et al. 2002).

These studies and the related literature have produced proposals for training that could anticipate the activation of subjects before starting both the rehabilitation sessions and those of study or training, or even as recharging exercises in moments of decay of sustained attention.

## 2.3  Executive Attention, Executive Functions, and Working Memory

After the presentation of five networks, we begin to understand that, if we also have to include the motivational and emotional systems, things get even more complicated. Therefore, in order to proceed with order in an orderly manner and to be able to explain the cognitive mechanisms—avoiding mere lists that require leaps of faith on the part of readers (with all the limiting consequences for the active professional) because they do not really explain—let's now investigate the mechanisms of executive attention and working memory. In this way, it will be possible to have greater autonomy and competence when using tasks, exercises, or treatments that involve the CEN.

Let's start with some considerations on executive attention. There are those who improperly divide attentive functions (focusing, selection, etc.) from executive functions (control, flexibility, etc.) and from working memory (maintaining and mentally reworking material on the spot). We prefer to refer to multicomponent systems that do not lead to the contradictions arising inside reductionist models, in which several authors say they measure different functions from each other, even if they use the same test (McCabe 2010; Benso 2018). Reductionist models are also widely used for their charm rather than for the representative strength of the neuroscientific reality. This happens because people want to improperly claim that with a single test, you can isolate a single executive function. Consequently, contradictions appear in literature that should make us reflect, but this does not seem to surprise us as much and we proceed improperly as if nothing had happened even if it is unscientific not to ask certain questions. In reality, you should know what is known in methodology lessons, that is, that the combination of a psychometric test with the function it should measure is not a fixed and certain operation but is due to the common sense of the experimenter and is a purely arbitrary step (Poldrack 2006). This arbitrariness remains despite this phase being then "covered" with attempts at legitimation (which start one step later) carried out with sophisticated statistical programs.[2] Already, Bernstein and Waber (2007) argued:

> The reductionist approach continues to influence neuroscience and in particular clinical neuropsychology. There is still the implicit assumption that functions are organized in discrete "packages", therefore it is inevitable to assume that executive functions can also be improperly analyzed as functionally discrete modules. From all this arises a further approximation that leads to creating test batteries that should discretely measure the different EFs that are divided into specific abilities arbitrarily linked to specific tests (e.g., Wisconsin Card Sort = set shifting).

For a clinician, it is important to also understand from the language used in conference reports or in the literature if he is receiving information that is not in line with the current visions of the brain models in neuroscience. The "suspicious indices" (D'Esposito and Postle 2015) are given by a terminology that should no longer be based on concepts such as "warehouses," complex encapsulated modules, simple areas that support specific functions represented by individual tests. The interpretations of the test results are not reliable either when they refer to models of systems that have already been historically modified. We repeat again that current knowledge is instead based on circuits made up of sets of multiple areas that can be more or less activated and that support multiple functions that are not easily isolated because they are very interactive. These circuits operate in close relation with other networks; they follow rules and precise shifts of intervention that are only now starting to be partially clarified (we have seen that the SNs pilot the intervention of the CENs and the DMNs).

In any case, it is good to never forget that uncertainty and doubt are the baggage of those who deal with neuroscience: Without these two components, there can be no progress with respect to the few fixed points that have been isolated, tested, and confirmed over the years.

Any clinician who works with a minimum of awareness can understand that a single test will never be able to isolate a single function, because to solve a task, the functions in action are always more than one, especially if the object of the measures is constituted by the executive attentional aspects.

In this regard, Rabbitt (1997) argues that "thinking of evaluating a single and hypothetical functional process and not others could be a completely inappropriate strategy to analyze an executive function, because an essential property of all 'executive' behavior is that, by its nature, it involves the simultaneous management of a variety of different functional processes." (Box 2.1).

---

[2]The use of factorial equations, although beginning to provide some correlational answer, is inserted only after the arbitrary act, which remains as an unavoidable starting choice. Such a choice is often not even supported by exploratory factorial equations that should legitimize the distributions of tests in the respective latent variables.

**Box 2.1 Some Suggestions to Avoid Superficial Evaluations of WM and Errors in Diagnoses**

For those who have experience with executive functions and working memory, it is natural to consider the importance of a battery to investigate EA and WMC, as also suggested by Hofmann et al. (2011). It may happen, and it is not a rare event, that in a battery made up of five tests that measure WM, a subject A may show weakness only in tests 2 and 4, while a subject B only in tests 3 and 5 and so on.

This is easily explainable if you evaluate the multicomponent nature of WM and the individual differences highlighted in many scientific works. In some subjects, idiosyncratically, some functional (executive) profiles could be underdeveloped or in others not yet brought into play for various reasons, including emotional motivational ones.

It is interesting to note how risky it is, from the point of view of centering the diagnosis, to administer few or single tests to evaluate this multicomponent set defined by the acronym WMC.

It would be enough to reflect on the fact that subject A would not show any fragility if he were subjected only to tests 3 and 5 and, if so, it would be really naive and misleading to write in the report: "He has no problems with working memory and executive attention." This is not such a safe ground on which to express oneself and this is perhaps clear only to experts in attention and working memory. In this case, precisely evaluating the need for a substantial battery and the possible scarcity of tests available, when it comes to working memory or executive attention, experts suggest not to dismiss the problem too soon. Such wrong practice is not so rare, especially if few WM evaluation tests are administered. Benso (2018) accurately illustrates this conduct, referring to the logical fallacy of "denial of the antecedent" or "improper commentary of the null hypothesis," as psychometricians would say.

In these cases, the advice of experts is to be cautious from the point of view of the method and to get used to not putting oneself in the risky situation resulting from the fact of providing, in diagnoses, information that would turn out to be false once examined in depth. It is better to report that the subject did not fail the few tests administered rather than incautiously formulate the challenging sentence: "He has no problems with attention or working memory."

For those with a basic neuropsychological training, these suggestions seem superfluous. A neuropsychologist always evaluates very carefully the data that emerge from the tests and, before interpreting them, compares them with theoretical models to evaluate their coherence. Normally, he does not completely trust the results obtained with few tests and even less of the subjectivity of the questionnaires. The judgments issued are tendentially conditional and probabilistic, and the professional expresses himself after the diagnostic inferences (from which, willingly or not, one must always start) are directed toward verified hypotheses that acquire a reasonable percentage of reliability,

(continued)

**Box 2.1** (continued)

also because they are validated by the functional architectures that one must adequately know.

On the other hand, those who study the models in depth, and know them very well, understand why Hofmann et al. (2011) recommend a battery of tests to have a certain sensitivity in isolating WM problems. In this sense, too many omissions (false negatives) seem to be scattered in diagnostic reports, especially in the developmental age.

Thus, Costa et al. (2014), regarding a single shifting test, state, "Shifting supports several complex mental operations, as the ability to cope with changing environmental conditions; it is known that it is a function impaired in people with cortical injuries (especially lateral frontal or orbitofrontal and subcortical cortex). However, it is impossible to make a pure evaluation of the 'flexibility' function because, to perform these tests, other abilities such as abstract reasoning, executive control, and working memory are necessary."

A way out of these conflicts consists in starting to understand that, from the point of view of methodological correctness, it is better to think of a more generic model that does not deny the existence of executive attention functions but that subsumes and incorporates them into a multicomponential system.

There are several similar models in the literature, one of which was proposed by authors who deal with working memory and is that of Executive Attention (EA) as intended, for example, by Engle and Kane (2004). It is a complex system that is expressed with working memory, which the authors call Working Memory Capacity (WMC). According to its theorists, such a system works with a concentration of attention resources useful for keeping the goal of the moment active, without being captured by distractors. It would contain all the executive attention functions in reciprocal interaction (useful for the different systems of re-elaboration and re-updating of the material). This interaction implies the more or less intense involvement of a group of functions, and this gives rise to the individual differences that we find in WM tests.

### 2.3.1  Working Memory, Same Terms, Different Models

On the expression "working memory" (Working Memory, WM), so widely used, different denotative values emerge in the neuroscientific models that want to represent it: We will now focus on this aspect for the fundamental importance that such a system would have for diagnosis, habilitation, and learning.

Working memory also has a very important history in the field of neuroscience, especially for the radical transformations that the first models have had. An authoritative pioneer, among the best known, is Alan Baddeley who, together with Hitch, in 1974 proposed a model of working memory among the most studied by the neuroscientific entourage. Not everyone knows that this model changed radically and developed further at least four times over the years, to currently merge with the one proposed by Engle and Kane (Hofmann et al. 2011). Baddeley and Hitch (2000) showed great intellectual honesty in the scientific field when, criticizing their own model, they highlighted its shortcomings; it is worth emphasizing, however, that in science in general, perfect models do not exist.

Paradigms serve to represent a reality that will never be the true one, as physics itself teaches us: The best model will be the one that provides more answers to the questions posed by the real world. However, the most appropriate choice is not always the most rational one but rather the one that is most shared by an entourage that has greater influence due to scientific or other merits, for example, because at a given moment, it has a particular media and academic weight. Science itself explains that an expert could be involuntarily influenced in decisions by emotional-visceral choices, mostly by the "trends" of the moment. None of us is completely immune and free from such forces, as demonstrated by three psychologists who won the Nobel Prize by demonstrating the validity of these disconcerting arguments. We refer to Simon, Nobel Prize in 1978, author of the definition of "bounded rationality," taken up in a certain sense by the other two Nobel winners, Kahneman, in 2002, and Thaler, in 2017.

For example, Cowan's working memory model, dating back to 1988, was already more complete in explaining the brain mechanisms than that of Baddeley and Hitch of 2000, but it began to be made known in neuroscience books only about 15 years later. Let's not forget, however, that even the primitive Baddeley and Hitch model, although poor in relation to some components, still led to important conquests especially in the field of the linguistic and visuospatial system.

The 1974 model and, later, the richer one of 1988 used explanations that were still terminologically based on encapsulated and not very interactive stores (especially with long-term information) and did not explain much about the re-elaboration of online material (re-updating and re-elaboration constitute one of the main functions of working memory). Things have changed significantly after 2000, but it should be remembered that among the most important models that have inspired various authors is the one already cited by Cowan (1988; Cowan et al. 2002), as also openly admitted by Engle and Kane (2004), who already at the end of the 1990s presented a very complete system (Engle et al. 1999).

It is good to know that the fundamental processes that constitute Cowan's working memory are purely attentional: The focus of attention is supported by the resources of an executive system that is internally projected toward long-term memory.

The interaction between the different circuits is the rule and the Executive Attention model that subsumes all executive attention functions (known and unknown, therefore not divisible is the most real in the light of neuroscientific evidence). The models of Baddeley and of Kane and Engle are currently overlapping because they provide common explanations of a multicomponent executive system as a nodal point of the WMC.

This overlapping of models was already intuited in 2006, when Repovš and Baddely described the central executive with these words (and note how many times the term "attention" appears): "The central executive seems heavily involved in complex cognitive abilities, acting as a sort of attentional control that initiates the focusing of attention, the division of attention between competing tasks, and has a component of attentive switching. In many of these functions, the central executive is supported by the other components of Working Memory" (Repovš and Baddeley 2006).

In Box 2.2, we will provide a very simplified description of working memory in order to communicate its fundamental importance in the various processes that occur at the brain level.

Let's start with a general rule that distinguishes short-term memory from working memory, which in turn, as we have seen, takes on further implications based on the period and the model to which it refers.

According to Benso (2018), short-term memory or Short Term Memory (sTM) is a system in which one enters and maintains a limited number of linear information for a short time, to return it as an output information without any transformation: $A - iA = A$ (where "i" is a sort of "identity function," useful for keeping the representation of the incoming information unchanged). Instead, working memory, or Working Memory (WM), is a system in which a limited number of linear information is input and maintained for a short time, to be output transformed or re-elaborated: $A - trA = trA$ (where "tr" is a function that transforms and reprocesses the information).

However, these definitions are very generic, precisely because of the diversity of the models that represent them and the multiple meanings that the same term has had over time.

**Box 2.2 A Discursive Description of Working Memory**
We can thus say, simplifying, that working memory or Working Memory Capacity (WMC), inserted in the Executive Attention model (Engle and Kane 2004), assumes a central function for learning and enabling processes. It intervenes whenever one must perform mental operations and reconfigurations that require maximum concentration. The executive attentive resources contained in it are limited and vary from individual to individual. When these resources are insufficient, difficulties emerge in reaching levels of expertise in various cognitive areas (especially memorization, expression, text comprehension, problem solving) and motor areas (learning of complex gestures). Another important function delegated to this attentive executive system, which is expressed with the WMC, is that of regulating emotions and behavior. A weakness in this sense could negatively affect performance, with the uncontained emergence of intrusive, devaluing, and anxiety-inducing thoughts, disturbing memory, and performance at the moment in different areas (school, sports, social life). However, it is worth noting that the set of such systems can find improvements in functionality, through expertly selected and targeted treatments and stimulations (cf. Takeuchi et al. 2010; Benso 2018; Benso et al. 2021).

Let's give an example in order to understand, trying to read for a few seconds and memorize the following five words: "sea," "dog," "sun," "screw," "blanket."

Afterward, you have to close your eyes and try to repeat them directly as they are written, then try to mentally put them in alphabetical order.

Then read them again once more (in the meantime you will have already done it!) and then try.

The first exercise is a short-term memory (MBT) test that, as D'Esposito et al. (2000) say, activates the DLPFC circuit (area 46) little or not at all while the second test, called Alpha Span, consistently activates this circuit, which is a hub of the CEN.

## 2.4   Neuroscientific Literature and Fundamental Rules for the Effectiveness of Interventions

In the studies of D'Esposito et al. (2000), it is clearly seen how only re-elaboration exercises activate the lateral prefrontal cortices. This valuable observation should be considered together with a work by Wagner et al. (1998) in which it is demonstrated that there is little memory of stimuli that are processed without the activation of the fronto-lateral cortices. These aspects have suggested to us the importance of keeping such circuits activated to allow learning to settle and, at the same time, have pushed us to avoid as much as possible intervention techniques that favor automatisms and stereotyped activities that do not stimulate the areas capable of supporting memory and learning.

Such precautions were already included in the work programs of the great masters of art and trainers. In confirmation of all this, Metzler Baddeley and colleagues (2016, 2018) document treatments for WM in which results are obtained on cognitive tests and also with "diffusion tensor" resonances. Two groups subjected to the treatment are considered: the first, defined as "adaptive," exercised with well-calibrated material in the sense that the software regulated the difficulty of the exercise on the subject's performance while the second group, called "nonadaptive," repeated slavishly and for the same number of hours exercises already learned and automated. The results were disturbing to say the least: The two groups stood out for the plastic changes in the brain structure and where in one group the thickness of the cortex increases and in the other, it decreases. If in one group there is an increase in fibers and myelin, in the other, a decrease is detected, an opposite plasticity. Even the results of executive cognitive and working memory tests showed improvements in the "adaptive" group and worsening in the "nonadaptive" group (Metzler-baddeley et al. 2016, 2018). These works have strengthened and confirmed what the great masters of the art of the past suggested.

Therefore, since the 2000s, the search for awareness during training and the abandonment as soon as possible of obsessive automatic repetition has been a strong point of the integrated cognitive treatment (Integrative Cognitive Therapy, ICT),[3]

---

[3] Integrated cognitive treatment is expressed in the Benso method (Benso 2018; Benso et al. 2021).

but the other important point was represented by the fact that, thanks to the above-mentioned literature, we have understood how to measure and how to stimulate these circuits so important for learning and enabling the proposed material to settle, allowing progress to be made in the skills to be developed. In fact, following the work of D'Esposito, Postle, Wagner, and others after the 2000s, the method has been equipped with screening tools, for example, through the creation of different tests to measure EA and the WMC. At the same time, treatments aimed at stimulating the CEN circuits emerged (always driven by the literature of those years or shortly after), N-Back exercises were used (remember back 1, 2...), with re-elaboration in working memory, with "dual tasks" with shifting activities (quick change of task), presented in many different ways both in terms of difficulty and type. One of the most powerful exercises that we propose and that we have learned from various masters of art and sport is that of visualization and management of the mental image: Many professional artists or athletes of complex sports have practiced these types of training that, in addition to improving performance, require sustained attention and enhance working memory.

The knowledge of the models and executive attention circuits that we have exposed in these pages leads us to make a reflection now: It is necessary to take into due account the innovations suggested by the different levels of interaction between the CEN, DMN, and SN circuits. In the sense that with what is known, detecting falls in executive tests and immediately thinking of a weakness in executive functions, or rather in the CEN, would be a "naive" conclusion. This was done (and continues to be done) when a certain weakness to executive attention, for a fall in an executive attention test, was "linearly" attributed.

Now, however, we need to ask ourselves more questions that lead to four possible hypotheses:

1. The weakness could directly affect the CEN and working memory.
2. The problem arises because the SN does not insert the CEN in due time.
3. The measured weakness depends on repeated and uncontrollable intrusions of the DMN with intrusive, depressive thoughts and ruminations.
4. The set of networks is unable to coordinate adaptively for the activation of an attention concentration system.

Therefore, given a fall in an executive test, it is necessary to reason before making a decision on the system actually involved to follow one of the four hypotheses. It is no longer as obvious as it seems to be and there is nothing left to say except to conclude this important excursus on executive attention systems and working memory. Let's now propose a useful challenge in the analysis of facts and events that come from everyday life with the cut of a story that accustoms us to interpret the theories of executive attention.

To do this, in a more natural and less academic way, we will enter in an environment often used by attention theorists to explain some processes: that of the cocktail party.

## 2.5    The Different Types of Attention Applied at a Cocktail Party

In this story that takes place during an executive attention cocktail party, we will introduce new types of attention, useful for our purposes, and they will not yet be all those that are known. We will also see a first application that leads to a brief excursus on motor learning that will be useful, together with a subsequent more in-depth explanation, to also define the treatment model of atypical swallowing.

Such an approach will also help operators to better internalize the "attentive" readings of the processes in progress, depending on their use in their own enabling sets.

To do all this, we will use a little story in which any coincidence with real facts or characters is purely coincidental. Let's imagine following some moments in the life of Mr. Tom C., a Navy employee with several sports passions as well as acting. At this moment, he is entering a large multifunctional complex, which he frequents regularly, where a summer party is taking place. It is the engagement party with the future fourth husband of a club member. This is Mrs. Lucrezia B., known for her passion for preparing wonderful cocktails and tasty pastries. Tom does not know and has never seen the birthday girl, who is not regular at the club, so at the entrance, he asks for information from the gentleman who checks the invitations and inquires about the appearance of Mrs. Lucrezia who, out of politeness, he would like to go and say hello first. The lady has been described with various attributes (average height, blonde, bobbed hair, blue jacket, white shirt, greenish skirt; for the reader who is sensitive to fashion combinations, we apologize in advance, also on behalf of the lady). Tom enters the garden and immediately notices a large swimming pool, at the edge of which cocktails are served, but his first task is to find his objective, a visual search guided by the attributes previously memorized and illustrated in Box 2.3, what attention experts call an attentional template. A cluster selection process is born to isolate the target from the distractors (selective attention). It should not be overlooked that everything depends on an adequate working memory to keep in mind the attributes that guide voluntary visual search.

Returning to our "attentive" story, there are other conditions to observe so that the operation is carried out optimally and in due time. First, sustained attention must remain active on the goal and any distractions: For example, a beautiful basket of fruit and sweets available on a table could suggest to Tom different gastronomic fantasies, interrupting the flow of the search.

Working memory is still crucial to keep active, sustained attention on maintaining the goal, despite the presence of numerous distractors.

**Box 2.3 Information on Visual Search**

Visual search is a very broad and interesting branch of attention. Here, we will only be able to make very brief references useful for our current purposes. This research can depend on the environment that, with certain stimuli, drives the subject's attention from bottom to top (it is called bottom-up) or it can occur voluntarily, according to internal strategies or motivations from top to bottom (it is called top-down). Some experts would say that the search mechanism would be that represented by a map (top-down) that is compared with that of the stimuli (bottom-up). Only when the two maps coincide will the search be considered completed. Based on the bottom-up maps, the selection style can also change. For example, it is easy to select a red dot among many green dots; it is so easy that it is said that the target being searched has a pop-out effect (it catches the eye). The effect is immediate and does not depend on the number of green dots (indeed, the more there are, the easier it may be to see the red). In the case of our example, it is instead like looking for a green vertical line in the middle of green horizontal and oblique lines and blue vertical lines. The search is slower than the previous case; it is almost serial (Treisman would say) or by groups of stimuli (chunk) with a more concentrated attentional focus Bravo and Nakajama (1992) would say. In this case, the number of distractors can influence the search times, which is a bit what happens to Tom in the search for Mrs. Lucrezia among so many people dressed in clothes of many different colors.

There are other prerequisites or underlying processes that help this cognitive operation. We must be physiologically activated in the optimal way for the maximum performance in visual search. Neither too excited, nor too sedated. It is the optimal *arousal* curve, that state of wakefulness that we recover every day after sleep, thanks to the activation of the ascending reticular system, as described by Moruzzi and Magoun (1949). The optimal arousal is given by a certain physiological activation that must not be low (microsleep or physical exhaustion) or high (excitement). In other words, the optimal performance respects Hebb or Yerkes and Dodson's "inverted U" law and the activation must be found in correspondence of the centrality of the inverted U.

Given equal physiological activation, what matters is the investment of the different types of attention that must be sustained and focused on the target. Meanwhile, Tom focuses attention on clusters of people to exclude or to consider someone who approaches the characteristics set in the template. As soon as he thinks he has found the target, he narrows the focus of attention to select the target among the distractors. Once he has assessed that it is not yet the right target, he reopens his attentional focus and sizes it on the optimal visual search. While he strategically shifts spatial attention (voluntary orientation), a firecracker that a group of playful customers suddenly explodes rightly distracts him. He is attracted by the sudden event that must be monitored. In such case, as previously explained, the system of automatic orientation, which is innate and guided by stimuli, can interrupt a voluntary process. However, Tom's will to pursue the goal, as soon as the capture effect is over, makes him refocus on the purpose of the search. Tom finally sees a lady at the edge of the pool (selects) who has all the attributes of the template: He meets her and introduces himself; indeed, she is Mrs. Lucrezia.

The search with the goal-centered template was very efficient, but some small typical inconveniences of excessively focused attention were detected, dimensionally closed on certain attributes. Among the guests was also Tom's brother, and a few minutes before, Tom himself passed in front of him and did not greet him because his attention did not "see" him; it was very focused and centered on attributes of a template that did not coincide at all with that of the brother (sex, hair color, and clothes). This can happen very often. Attention is a difficult subject to deal with: If you are very efficient in the "local" dimension, you lose the information of the "global" dimension. If you are very efficient in the global, you lose the details of the local; the capacity of resources, in fact, is always the same. Obviously, Tom and his brother do not know these things since they do not experiment on attentional theories, and therefore, the fact was unpleasant to both of them. These are attentional phenomena on which shows like candid cameras are based, a form of change blindness or "blindness to change" that occurs when a detail in a scene changes under everyone's gaze, but no one notices it (Turatto 2000). Now, Tom is talking with Mrs. Lucrezia, and in the buzz of the party, he has to use selective auditory attention to isolate the vocal messages. In the meantime, someone nearby says his name and this "powerful stimulus" penetrates the attentional filter and enters his conscious perception. Tom manages to keep his gaze focused on his interlocutor; however, for a few seconds, he is interested in what his neighbors are saying about him: He shifts his attention without moving his eyes, while the opposite procedure is not possible.

*Tom knows Dorian and trains his phasic and tonic alertness on the tennis court.* The evening gets long and to find a reason to stay, games are organized on an adjacent illuminated tennis court, with a machine that shoots balls at adjustable speeds. Mrs. Lucrezia's future husband is called Dorian G.; he looks very youthful despite his age and is a passionate collector of paintings. Dorian has been training tennis players for years and Tom, who plays as an amateur, wants to try the ball-shooting machine. Dorian is willing to calibrate the speed of the balls to Tom's response ability and does it very well. The machine emits a warning sound and, after a variable interval of a few seconds, shoots the ball. Attention experts would say that between "ready" and "go," there is a response preparation curve (a "U" shaped warning curve as we have described) and this short interval is defined precisely as "phasic alert" which, if it lasts longer, becomes "tonic."

Tom activates himself attentively to respond in time and a sequence of balls has transformed the session into a sum of phasic alerts that over time, becoming tonic, transform the whole thing into an intense sustained attention session lasting about 20 min. Tom feels particularly good after this exercise: Also thanks to Dorian who has perfectly adjusted the speeds, calibrating them to Tom's degree of expertise.

During this time interval, different aspects of attention have been brought into play (in this case, tennis). Various attentional aspects have been touched (a fair arousal is assumed, in the sense that Tom must not have overdone it with the alcoholic cocktails): voluntary and automatic orientation; phasic and tonic alertness; focused, selective, and sustained attention; control functions; flexibility; emotional self-regulation; and therefore working memory, involved in all these activities. Tom is doing well and has improved not only his tennis skills but also his various attention networks, thanks to Dorian who graduated the speed and who has maintained an activation time (20 min calibrated exclusively for Tom) that has not come to heavily stress the attention system like an entire match which, with emotional implications, must find athletes very gifted in attention resources and well structured on the self-regulation system (this applies to any treatment).

The inexperienced player must control the motor act not yet learned with the CEN. A less experienced tennis guest than Tom, a certain Robin H., an excellent performer and instructor in archery, but who has little tennis practice, also wants to try the ball-shooting machine. In this case, even if the shooting speed of the machine is lowered, Robin must review some motor calibrations. Let's just say that the movements of archery do not facilitate and interfere with those necessary to manage a tennis racket. As Buccino et al. (2004) and Vogt and Magnussen (2007) would say, the mirror neuron system of this subject would be much more activated with the gestures inherent to archery, compared to the movements of tennis. Therefore, not having the fluidity of the tennis expert, Robin must break down, assemble, and recheck the gesture with the executive attention system in action. In other words, the CEN networks including working memory intervene to support the learning of the gesture. Instead, an expert in this task unloads the control systems having refined the movement and performs the motor act with ease, without involving the CEN except in case of particular unforeseen events or changes in trajectory.

Tom loses the match due to performance anxiety that causes an excess of control. A regular visitor to the club, a certain Gioachino R., owner of a copy shop and avid fan of opera music, challenges Tom to play a single set of six games. Tom accepts because he has some unfinished business to settle. He has lost to Gioachino several times and suffers particularly from him, also because he does not bond much from a relational point of view.

Tom finds a worthy opponent who challenges him on every point and begins to feel increasingly tense. The more the match approaches the end, the more the emotional tension rises and the arm becomes less and less flexible because the points now weigh and the fear of making mistakes increases (therefore trivial errors increase). This is what commentators call "the tennis player's dead arm" (Box 2.4): Too much control leads to gross errors even for a super professional. Neurofunctionally, we can say that the emotional systems that have privileged connections through the amygdala with the SNs send information of anxiety and fear of making mistakes. The SNs in turn dysfunctionally ask for help from the control system, which they insert too massively. What was a well-trained fluid gesture in handling the racket becomes a hyper-controlled, slowed-down gesture.

**Box 2.4 Tennis Player "Dead-Arm" and Stuttering**
Many of the examples we give, using sports activities for clarity and greater simplicity, are transferable to the world of school and clinic: This has been understood for years by skilled rehabilitators who have retranslated the general models into specific activities for their clinical, educational, or sports skills. Deep experts in single disciplines, who discovered these models of brain network functioning, almost immediately saw the connection with their specific situation, in the clinic, in the classrooms, or in other realities. As an example, we cite some stuttering experts who were already using effective treatments for this dysfunction. They needed a model to explain the mechanism of what they were obtaining with some subjects suffering from stuttering. They explain that the excessive activation of the emotional system seems to insert the CENs in an overly controlling way, as in "tennis player's dead arm," and these can end up blocking or increasingly hindering the fluency of speech. The use of double tasks well calibrated on the subjects and rhythmic (not a random task) improved verbal fluency in these subjects. Some of these experts (Doctors D'Ambrosio, Di Maro, Di Somma) expressed the point of view according to which a double task calibrated on the subject could discharge many of the control resources, partially freeing the linguistic act which regained greater fluency. It should be noted that we are not talking about any double task but we said: "calibrated on the subject and rhythmic" (D'Ambrosio et al. 2016).

Indeed, a hyper-learned gesture, when it comes back under control systems, is inevitably slowed down and loses the fluency acquired after hours or years of training.

*Tom loses the match because he can't remain centered in the attention set. Similar aspects can be found in school and in the clinic.* Tom loses the match. He was psychologically tired and a bit tense compared to Mr. Gioachino: In a moment of reflection, he also notes how in the moments of pause between the sets he did not manage the flow of thoughts properly. The management of attention even during pauses is fundamental in various activities and especially in tennis. Many teachers or coaches are focused on the phases of performance, but the results are also built during the intervals and pauses. The amateur coach is the one who only knows how to give advice like "Don't think about the mistake!," but in this way, you end up thinking even more about what, instead, you shouldn't be thinking about. First, you have to prepare the athletes to keep a thought centered even on a secondary task or on a ritual that does not stress attention too much but that maintains awareness in the set with a minimum of effort. This also applies to kids who suffer from anxiety and get stuck on oral or written exams. Recommendations to apply at the time of the task are not necessary. They must be prepared beforehand with adequate breathing, with anchors to thoughts that follow well-defined logics. They must be desensitized and trained with these techniques for a few minutes, but almost daily, so as not to arrive unprepared at the moment of the anxiety-inducing event. In other words, you cannot improvise control and self-regulation techniques when you are exposed to the stressful event, but you must prepare them and anticipate the moment of the most intense emotional flow, with desensitization also experienced through certain types of breathing, to start several minutes before entering the environment of the event to be faced.

All this also applies to learning situations, and in our case, we are talking about atypical swallowing: The distracted subject, who is worried and not centered on the task, delays the learning and enabling processes.

Let's go back to tennis in order to make the story easier, but we invite you to understand the models and metaphors to be transferred to the professional field you want to consider: You should not let your thoughts run wild after serious errors or, even worse, in cases of hasty convictions of victory.

**Box 2.5 The Control of Intrusive Thoughts Through Rituals, as Taught by Rafa Nadal (A Kind of Dynamic Mindfulness)**
We have just illustrated some possible causes of a difficult interaction between the networks, due to a lack or excess of activity of the orthosympathetic system. Now, however, let's hear how the attention set is self-regulated and maintained during the "emotional storms" of a tennis match that can last 3 or 4 h. His effectiveness and his performance were there for all to see, but few understood the value of what they were seeing: Rafa Nadal's behaviors were often judged superficially and misleadingly by those who could not have known what is being explained here, but that Nadal and his entourage empirically knew very well. It should not be overlooked, right from the start that we are talking about the demonstration of one of the best trained and balanced cognitive systems in the world of sports in general. Nadal was instead accused of obsessive-compulsive and tic behaviors. In our small scientific laboratory bunker, this doubt never arose: We dismantled this gross clinical nonsense by saying that if it had been so, he would have had to transfer these behaviors, which by nature transcend the will of the individual, also in daily activities and therefore during interviews, walks, or moments of stress. This is not the case because the gestures and techniques are neither obsessive nor superstitious. They are conscious actions that follow a logic that keeps the thought "here and now" without allowing major intrusions by the DMN in the most delicate or difficult situations of the match; a statistic, in fact, shows that in the most important points for the outcome of the match, Nadal's success percentages are very high. It is a sort of tea ceremony, a "ritual" that calms and keeps the subject anchored to the present moment: Let us remember that anxiety arises when you project yourself toward the future or the past. The "here now" is free from any such feeling.

(continued)

**Box 2.5** (continued)

Let's now try to discover Nadal's true intention, definitively clarified in 2020 with an interview given to the "Corriere della Sera," of which we report a short extract functional to our purposes.

Interview with Nadal.

from the "Corriere della Sera" (November 1, 2020).

People wonder the reason for his ritual: the two sips of water from two bottles, the lines not to step on... Superstition?

No. I'm not superstitious; otherwise, I would change my ritual after every defeat. I'm not a slave to routine either: My life is constantly changing, always on the move, and competing is very different from training. What you call tics are a way of putting my thoughts in order, for me who am normally very messy. They are a way for me to concentrate and silence the voices inside. To not listen either to the voice that tells me I will lose, or to the even more dangerous one that tells me I will win.

But when in the locker room you put on your bandana and shout "Vamos!" You are the one who scares the opponents.

I don't shout in the locker room! I take a cold shower, listen to music on my headphones, and, yes, tie my bandana. But I have never allowed myself to intimidate an opponent.

Very interesting is that even more dangerous voice "that tells me I will win": It's a way to let go, let the mind wander, step out of the attentive set, and then have difficulty getting back into the right mood, caught in the panic of a possible comeback.

During the preparation of the shots or during the change of field, it would be good to maintain a minimum attention on particular rituals that every athlete must be able to train, so as not to let thinking escape toward the most negative part of the DMN. Tom that evening was losing concentration also because he often entered the DMN, caught by intrusive thoughts like the one that caused him the displeasure of not having said goodbye to his brother; from that thought, it was then natural to move on to other "depressive hooks" to get to performance anxiety. In other words, Tom, driven by emotional impulses, enters the DMN when he should concentrate and then can no longer find the attention state necessary to "be in the game." At this point, he can do nothing else, as said before, but let himself be guided by emotions and try to insert the control system in a massive way, out of time and out of calibration with the consequences that fall on game strategies, motor coordination, effectiveness, and accuracy of the shots. To better understand what we are saying, we report the opinion of some scholars on this phenomenon due to excess emotionality in Box 2.6. Instead, in Box 2.5, we saw a positive example of an athlete known as one of the most self-regulated and cognitively powerful: Rafa Nadal.

*Excessive emotionality and very high stress invert the expected adaptive situations.* Starting from the survival of the species, we can say that a correct emotion is one that inserts the CEN to control the situation, if there is a need. For example, the motor act of walking normally, freely on easy surfaces, rightly changes state when we find ourselves on a dangerous mountain path—in this case, in fact, we carefully look where to put our feet—or when we have to cross a busy road without a traffic light and we have to evaluate the speed of cars, spaces, times, and directions of travel. These are phenomena of "correct" intervention, dictated by emotional sensitivity that captures the danger, an adaptive phenomenon for the protection of the species. However, if in these vitally important cases emotionality invades the salience networks (Box 2.6) and blocks the state change systems, making them unstable, there is a risk both of inappropriately inserting and disinserting CEN and DMN and of remaining at the mercy of the most instinctive and orthosympathetic subcortical processes.

**Box 2.6 Emotionality Excess and Attentional Networks, What Emerges from the Scientific Literature**

We report what some authors maintain in the interaction between emotional and cognitive: According to Eckert et al. (2009), their study could highlight the extent to which activity in the AI/FO (anterior insula and frontal operculum) precedes the onset of DLPFC (causal effect of the SN in the insertion of the CEN) related (Sridharan et al. 2008) and the extent of the activity degree of the right AI/FO. They suggest that the ventral attention system modulates the excitability of the dorsal attention system and the specific task-related systems. Too little activity of the right AI/FO or too little arousal in the autonomic system, that is, attention deficit (Dickstein et al. 2006), may fail to insert the DLPFC, thus determining neglect errors, while too much activity of the right AI/FO limits the DLPFC function and the selection of optimal responses in people with high anxiety (Etkin and Wager 2007).

One can remain blocked, without being able to move forward or backward on the mountain path or while crossing a street, in a sort of freezing or cognitive motor block, or perform dangerous indiscriminate actions (attack or flight), running or moving indiscriminately in dangerous situations.

*Tom makes some reflections.* Although he does not know cognitive neuroscience, Tom is aware of how he could have lost the match because some mechanisms that prevented him from maintaining sustained attention on the task failed. He can understand how difficult it is to get back into the set when chasing thoughts dictated by the emotion of the moment. He can make these reflections also, thanks to drama lessons, in which the interpretation of various characters leads many actors to become aware and introspectively conscious of the different mental states with which human beings operate. Furthermore, having to often confer with radar operators as a Navy employee, he also understands their type of attention, still different from those encountered so far. Attention to rare stimuli is a rather difficult exercise that requires different training: When we monitor an environment if nothing happens, we get distracted much more easily and our thoughts take flight in their innumerable chains; distraction in DMN is the rule. Therefore, when something suddenly appears, someone who is wandering in his own thoughts might not notice it or realize it too late. The radar operator's job is precisely to intercept as soon as possible any variation from the preordained routine (often monotonous and repetitive): This is the function of "vigilance" which differs from sustained attention, because the latter follows the event with continuous responses while vigilance is tested with rare stimuli. Both need resources and are maintained longer over time if the subjects are highly motivated and positively charged on an emotional level.

*Tom in the face of complex motor learning.* Tom's following activity of will introduce us to important aspects of motor learning, because it is this knowledge along with attentional knowledge that we should have in order to tackle enabling techniques, including the improvement of swallowing. After all these activities, it is now midnight. Many guests have left the center, but there's still a small group that, to the sound of music, is dedicated to learning the steps of a new dance. There is a highly rated teacher, a certain Fred A., who used to work as an employee clerk but now only deals with performances and dance lessons.

Tom sees that his brother is there too and joins the group. The gesture they are repeating to the rhythm of music is very complex. Tom immediately feels clumsy and uncomfortable but little by little begins to sketch the gesture. He helps himself by breaking it down into many sub-movements, even verbally repeating what he sees and has to do ("left foot forward, right arm up, right foot forward and equals left foot, both arms up"). He breaks it down, slows down, then picks up the rhythm of the others, then stops, repeats a sequence that he kept getting wrong, and then starts again. Fred sees him, approaches him, and makes him repeat the sequence, acting as a slowed model to slowly get him to follow the rhythm of the music in time. Now Tom, if he is not always distracted with the CEN activated, performs the gesture in time. His movement follows all the expected steps, but even though it is learned, it still differs from Fred's because it needs to consolidate in memory to 1 day be able to do it with the necessary expertise that discharges the CEN. We would say that the resources (of limited capacity) that Fred saves in producing the hyper-learned gesture can be used in the artistic expression of the gesture and in consciously feeling the sublime sensation of a movement that merges with the music and that expresses itself as the art of dance would like it to be: through body, mind, spirit, and sacredness of the gesture. Until the attention networks necessary for the formation of the complex gesture are transcended, it is not possible to reach certain peaks of artistic elevation. But let's now look at the most neuroscientific part that concerns us in this area referring to swallowing.

Tom concludes the evening. Fred's dance lessons are coming to an end. By now, all the remaining guests are tired and a little nervous. Someone thinks about getting a ride home, not trusting their own driving, both because of the alcohol level and of the intake of products that cause drowsiness or attention lapses not always known.[4] This, however, is another story that helps to understand how "attention" is not the obvious term that many believe they mean. Obviously, this would open a further thread that we will discuss in another context. Now, we say goodbye to Tom, Lucrezia who collects some delicious leftover pastries, Dorian who thinks about his paintings in the vault, Fred and Gioachino who set off together in the car singing and walking to the rhythm of music, and all the characters who allowed us to review different aspects of attention, thanks to their different inclinations.

## 2.6   Complex Motor Learning

Motor activity involves many circuits also involved in cognitive problem solving. Therefore, as we will see, these distinctions between noble and less noble subjects, discriminated against even at a school level, if evaluated on a neuroscientific basis would not exist and indeed one strengthens the other and vice versa. This enormous possibility of growth, which explains the pleasure of movement and artistic activity, does not seem to be known to those who should decide or advise the drafting of school programs.

Let's leave these issues aside, convinced that it takes more time to get rid of certain stereotypes and outdated beliefs, and let's delve into what is our task: to operationally clarify the relationship between motor and cognitive.

Adele Diamond already in the year 2000 titled *Close Interrelation of Motor Development and Cognitive Development and of the Cerebellum and Prefrontal Cortex*: In this article, the author argued that the cerebellum would not only serve the motor function but would also have a role in cognition.

On the contrary, the prefrontal cortex, through its connections with cortical and subcortical centers important for the control of movement, may have a role in motor function, not simply in cognition.

By now, hundreds and hundreds of works trace a relationship between cognitive and motor systems, for example, Sakai et al. (2002), consistently with what has been stated so far, isolate particular activations in the CEN circuit of the DLPFC (area 46) during motor learning.

Seidler et al. (2012) find a strong correlation between motor learning and working memory. Leisman et al. (2016) suggest that, although they have been historically seen as separate functions, it can be argued that motor and complex cognition are functionally connected and both have evolved in parallel, in an interdependent way.

According to Llinás et al. (2001), the oscillations of neural activities can represent both motor and cognitive processes, suggesting that both processes may share similar evolutionary roots. Similarly, the basal ganglia are argued to have

---

[4]For example, Marrocco and Davidson (2000) present a review of works with reaction times that use as an independent variable the chemical manipulation of specific neurotransmitters, each of which affects a particular type of attention.

corticostriatal circuits that serve both motor and cognitive processes (Middleton and Strick 2000).

Cognitive processes have become more refined because they are associated with the need to adapt more complex movements (Melillo and Leisman 2009; Leisman et al. 2014) that are probably related to the development of fine motor movements. Benso (2018) demonstrated the enhancement of executive attention and working memory after a full immersion in a 15-day tennis camp. As regards the clinical aspect, "cognitive motor," in the developmental age, we cannot forget the significant boost at a national level, and not only, provided by the studies of Professor Letizia Sabbadini, thanks to which there has been a general clinical progress, especially in the classification and treatment of linguistic aspects and dyspraxia (Sabbadini 2013, 2021).

## 2.7  Interactions Between the Mirror System and the Attentional Control Circuits

Mirror neurons interact with the dorsolateral prefrontal cortex and therefore with working memory in learning a new gesture. There are interesting observations made by Buccino et al. (2004) and by Vogt and Magnussen (2007), which take us to the heart and mechanisms of interaction between the mirror neurons system and the attention control circuits (CEN and DLPFC). It is discovered and confirmed that mirror neurons are activated above all by the vision of known gestures. Capoeira and classical dance dancers are tested and greater activations are measured when a subject looks at the specialty that is his/hers. The authors then also illustrate some mechanisms for learning complex unknown motor acts, as happened to Tom when he found himself studying dance steps. The authors then also illustrate that the complex gesture must be broken down into several simple sub-gestures, with a consequent strong activation of the DLPFC and therefore of working memory. The contribution and importance of working memory also for the visualization of the gesture and the maintenance of the motor image is also grafted in. At this point, the mirror neurons, which were not active seeing the unknown movement, recognize the broken-down gestures that are in the baggage of the "gesture lexicon." The next operation is given by the recomposition, as when Tom followed Fred, who showed him the movement in real time and slowed down. During the recomposition, the DLPFC is very activated: It must keep in mind the different motor acts to recompose them into the single action to learn! However, when the fusion occurs, expertise has not yet been reached because the gesture must still be controlled by the CEN. When you become truly expert, like Fred after several years of dance practice, the CEN are discharged: This is one of the most delicate phases of swallowing enablement, since the learned gesture must be consolidated and maintained as a primary process compared to before. As also supported by other authors (Jansma et al. 2001; Landau et al. 2004; Kelly and Garavan 2005; Benso et al. 2021), working memory and the areas of the CEN circuits (especially the DLPFC) are highly activated during the formation of learning and of complex motor skills; however, when expertise emerges, such areas are deactivated so that the system can express itself with the necessary fluidity.

## 2.8 Important Reflection for the Treatment Model of Motor Skills and Swallowing

Working memory acquires a fundamental role for learning in general and in this case must have enough resources to be able to manage multiple stimuli. If there are many motor acts to be recomposed and if the WM is fragile, certain types of learning will never reach expertise or will reach it after a very long time. It therefore becomes a priority to measure its efficiency and insert in the method specific exercises to strengthen it. The literature is rich in works that demonstrate both cognitive and structural enhancement (fibers and brain areas) of the circuits that support the operations of the WM. Such improvements, also demonstrated with the support of neuroimaging, free us from the skepticism of those who did not believe that the WM could be enhanced (Olesen et al. 2004; Takeuchi et al. 2010; Schwaighofer et al. 2015; Ciarmiello et al. 2015—here is the application of our integrated method on amnesic Mild Cognitive Impairment patients with PET findings; Metzler-baddeley et al. 2016, 2018; Benso and Benso 2023). The use of dual tasks itself becomes useful for these purposes for several reasons that we will illustrate below.

## 2.9 The Dual Tasks

The expression "dual task" indicates the addition of one or more exercises during a preestablished performance. It is an "umbrella" definition, in the sense that many applications with different purposes fall under the same name. For a very general illustration, we can cite Tara et al. (2015): "We propose that dual tasking is the concurrent execution of two tasks that can be performed independently, measured separately and with distinct objectives." There are many misunderstandings when introducing this category into the world of research and clinical practice, mainly because the common thought is that it is sufficient to add any task to an already given one to create a dual-task treatment. This is not the case: The additional task must be carefully chosen based on the difficulty gradient, the typology for a good match, and the purposes that can be very different, as we will see.

The easiest and perhaps most used example in books that talks about attention, to exemplify dual-task interference, is that of a person who drives and converses at the same time. These two actions are assumed to be sufficiently automated, so much so that entire stretches of road can be covered, even if very well known, without remembering at all what happened during the journey (e.g., if we stopped or passed with the green through all traffic lights): This is favored by a rich and emotionally engaging conversation.

However, when traffic becomes heavy or we have to proceed through an unfamiliar junction—or when one of the two tasks requires a lot of resources—we interrupt the conversation to only deal with the direction to take, due to the limitation of the resources available. This example is in line with the rules of the dual task because the subject, in this case, is able to perform the two single tasks well separately. The theories of the interference that a dual task can create are different and there is no agreement among scholars: For some, there would be a competition for attentional resources; in other terms, it would be a limited capacity model in the sense that, when people perform two nonautomated tasks simultaneously, the (limited) resources must be redistributed between the tasks.

For others (Pashler 1994), the "bottleneck" model is preferable, which is based on the idea that certain critical tasks must be performed in sequence and not in parallel, so a bottleneck arises at the level of central processing and choice of response. However, the second model also provides for a sort of limited capacity of resources, for the response processing phase, depending on the central systems in common. The task that will be chosen by the subject as the main one will proceed normally (in reality with some small interference), while the secondary one will begin its waiting phase called "psychological refractory period" (PRP).[5] The "bottleneck" theory would explain the formation of the PRP: Until the processing of the response of the first task is not finished, that of the second cannot begin.

There are essentially two hypotheses on how two tasks are carried out when they are slightly complicated: Either we carry them out serially, quickly jumping from one task to the other, and thus the slowdowns during the process are understood, or we carry them out in parallel and the decrease is due to the fact that one of the tasks uses part of the resources leaving less for the other, which is therefore slowed down. The evidence in the literature would suggest that largely automated tasks can be performed in parallel. Pianists or typists have been observed who, while playing or writing, are able to repeat words simultaneously without having drops in the two performances (Pashler 1998). Tasks that need several processing stages could be performed in series, as happens with a computer that, through rapid jumps in succession from one task to another, simulates the simultaneity of processing, or in parallel with the distribution resources with limited capacity.

The double task chosen is not always the optimal or useful one for that specific purpose: As we will illustrate, there can be dual tasks that destroy the performances instead of helping their development, if they are not well calibrated on the subject; let's see in the meantime the variability and the different uses.

A dual task can be used to achieve seven different purposes:

1. *Confirm* the automatism that has occurred, always appropriately calibrated: A task that reaches expertise is little disturbed by a concomitant dual task. The

---

[5] The "waiting" time necessary to finish processing the second task is precisely defined as PRP, precisely because it resembles on a higher scale the absolute refractoriness (closure of sodium channels) and relative (opening of potassium channels) of a neuron that has just emitted an action potential.

cited works, with pianists and typists, indicate precisely this useful aspect that can become a sort of test/measure of expertise not yet reached, if the two tasks showed significant increases when carried out simultaneously compared to when they are assessed individually.

2. *Measure* the resources that are at stake in the various phases of a main task: Following what was said in point 1, differences in cognitive commitment between different tasks or within the tasks themselves can be seen. In other terms, with the rules of the double task and the assumption of limited capacity, it is possible to investigate the resources or processes by manipulating the primary exercise, used as an independent variable that acts on a secondary task, which would thus become a dependent variable. We will give an example in Box 2.7, illustrating a work by Castiello and Umiltà (1988).

3. *Strengthen* the executive attention systems, bringing the executive attention networks to the maximum level of effort that the subject can bear with a calibrated double task: It is a training at the limits of the subject's potential that induces development and enhancement of attention resources.

4. *Promote* learning and achieving of a specific expertise, maintaining an optimal state of concentration during practice to promote the attentional awareness necessary for learning the primary task: As evidenced by literature reviews, in general, treatments carried out by inserting a double task obtain better and earlier results than treatments focused on the single task (Varela-Vásquez et al. 2020).

5. *Deconstructing* automatisms: When learning reaches very high levels of expertise, it may become locked in a rigid schema that is then unchangeable: It is suggested to deconstruct also through slowing down during training or therapy so that you can intervene to insert new refinements in progression. In this regard, to outline the considerable importance of what has just been stated, we recall that Metzler-Baddeley et al. (2016, 2018) find that stereotyped automated learning after a brief initial increase worsens the cognitive performance of subjects and creates a brain plasticity, assessed with diffusion tensor imaging and fractional anisotropy,[6] opposite to that obtained with adaptive learning calibrated on the subject.

6. *Unloading* the excessive control of the executive on the main task can occur when anxious phenomena are inserted through the SN (and related networks) and the CEN to obtain an overly pressing control, which ends up blocking or hindering the fluent expression of the module (Eckert et al. 2009; Box 2.6). Any performance in progress, even well learned, is slowed down to favor the monitoring required by hyper-activated and worried emotional system. The additional task would have the dual function of unloading part of the excess of controlling

---

[6] Fractional anisotropy (FA) is an important index that measures the integrity of white matter fibers. It can account for the brain circuits that are formed or that are being formed. In general, a high FA value is correlated with improvement in performance. The alteration of FA can originate from various factors that depend on changes in myelination, from density and integrity of axons and their membrane, from their diameter and more.

and restraining resources and of distracting the anxiety-inducing thought, the subject having to introduce an additional activity to attend to. This is the case of the "tennis player's dead arm" (with the athlete who, for fear of making a mistake, controls the overlearned gesture too much, which therefore loses fluidity) or of some types of stuttering (Box 2.7).

7. *Studying* the commitment of the three phases of performing a cognitive task (input, central reprocessing, output), as occurs in the models of Pashler (2000), evaluating the psychological refractory period: Given two tasks, the input and output phases, as previously mentioned, can take place in parallel, but according to Pashler, this cannot happen with the response selection phase that would encounter a "bottleneck." There is a waiting period called "psychological refractory period" (PRP; Pashler 1998) during which the second task is "queued" until the first (chosen by the subject) does not, in fact, overcome the cognitive phases of processing the material, to at least get to the selection of the type of response. Obviously, those who develop few resources or use them poorly, such as, for example, subjects with ADHD, show very long refractory periods, or risk skipping both tasks; in this regard, it should be noted that there are complex, but practicable, methods suitable to measure the PRP (Benso 2004a; Pashler 1998; Benso, Benso, in preparation).

We believe that this brief list gives a minimal idea of how little is known about dual tasks, mostly understood as a simple combination of two exercises. They are instead very interesting tools for strengthening, for deconstructing pernicious automatisms, and for investigating the processes.

Reviewing the literature related to dual tasks, very interesting aspects emerge that we thought of applying with the ABAED® method to swallowing processes. Dual tasks facilitate the learning of motor activities and therefore both training and rehabilitation.

Some motor activities that are widely practiced in life, such as walking, reach very high levels of expertise for all of us but are never completely automated; therefore, the system is penetrable by top-down processes (the control of the CEN on how we move in situations where there is a well-known risk of falling). We can say that it is as if there were a sort of "impedance adapter" (Benso 2004a) that allows a connection with the central systems in case of need. It's as if a form of implicit distributed attention warns of changes in context or the sudden appearance of new stimuli, even in situations that are not particularly active, a kind of implicit vigilance that systematically informs the SNs of the internal state and the context where both analyzers and effectors operate (Benso 2018), from which they would be repeated syncretically for each individual, with scan times at a certain idiosyncratic frequency.

**Box 2.7 Two Studies to Explain Some Uses of Dual Tasks in Research**

To provide practical examples of the use of dual tasks in research, we will briefly illustrate two works: The first study refers to point 2 of the previous list, when we indicate the use of dual tasking as an evaluation tool, useful for measuring the resources that are in play in the various phases of a primary task.

We present a work by Castiello and Umiltà (1988) in which the resources used in the different phases of the game of volleyball are studied. The subject is equipped with a helmet containing headphones and a microphone. The (secondary) task is saying "hop" as quickly as possible as soon as a short sound is emitted from the headphones. At first, reaction times for this task were measured when performed alone. Then the athletes entered the field and were engaged in a series of serves and receptions (primary task). Some specific phases of the receiver's "wait" were evaluated (the moment of the serve, the moment in which the ball passes over the net, the moment of the actual contact with the ball). Delays in responding "hop" emerged, compared to the times measured with the single task (control condition). These delays during the various phases made it clear that the greater difficulty (longer times to respond "hop") corresponded to certain different types of serve and especially to the moment of impact with the ball.

The second example instead illustrates the need for calibration and optimal choice of a dual task as in point 6 of the previous list, where reference is made to the treatment of stuttering.

In the case of stuttering, there is evidence in the literature (D'Ambrosio et al. 2016) of improved fluency using well-calibrated, coordinated double tasks. The added task has the function of discharging the excessive control of the CEN produced by the emotion that blocks fluency. Furthermore, there are fewer resources left to produce thoughts of counterproductive performance anxiety.

In this regard, we can say again that not just any additional task is worth it, but exercises must be studied to be combined to the primary activity. In the case of stuttering, for example, D'Ambrosio et al. (2016) promote the following experiment: They isolate 18 subjects with stuttering and subject them to three different conditions that we will call A, B, C. The three conditions rotate in order of presentation with the scheme of a Latin square where each can be presented as first, second, and third.

In condition A, the subject produces a monologue of at least 55 words presenting a previously read passage without any other associated activity. In condition B, the subject produces a monologue of at least 55 words presenting the previously read passage and simultaneously performs a complex and non-coordinated motor activity (shuffling a deck of 55 poker cards with forcedly asymmetric gestures). In condition C, the subject produces a monologue of at least 55 words presenting previously read passage, simultaneously and in a coordinated way with a rhythmic motor activity (flipping through the deck of

(continued)

**Box 2.7** (continued)

poker cards at the rate of one card for each word spoken). In condition C, CENs were discharged from language and were mainly engaged in managing the assembly of the two tasks into a single scheme.

The results were as follows: Condition A (single task) demonstrated 14.28 disfluencies.

In condition B, the subjects did not significantly improve (11.61 disfluencies).

In condition C, the subjects improve in a statistically significant way with 3.39 disfluencies.

This example makes us think, in addition to the usefulness of the double task to intervene on subjects with stuttering, also about the fact that "any" task is not always suitable in this type of intervention.

These fascinating aspects, which open new perspectives in the field of neuroscientific knowledge and also in the habilitation field, are still unclear, as Uddin (2014) states, but it is what we intend to continue to investigate in our study and research projects.

# The Cognitive Motor Reeducation of Atypical Swallowing

The transition from infantile to adult swallowing is gradual and occupies a long period in the ontogenetic development of swallowing. This phase is characterized by the coexistence of both infantile and adult-type swallowing acts in variable proportions, which tends to the progressive decrease of the infantile-type swallowing acts. The transition from one type of swallowing to the other should occur around 18–20 months (Schindler 1990), but around the age of 7, the copresence of the two modes of carrying out the oral phase is still considered physiological, no longer deemed acceptable with the eruption of permanent dentition. At the age of 10, 70% of subjects present a definitive adult-type swallowing pattern (Cozza et al. 2020), although some authors consider a lingual thrust swallowing physiological at this age (Hanson and Cohen 1973; Proffit and Mason 1975; Speidel et al. 1972).

## 3.1  Atypical Swallowing

The presence of infantile-type swallowing at an inappropriate age is defined as "atypical swallowing" (Garliner 1996; Lanteri and Lanteri 2001; Cozza et al. 2020), "deviant" (Lanteri et al. 2011), "deviated" (Andretta 2012), reverse (Straub 1960), or "dysfunctional" (Polimeni 2020; Schindler 2020; Levrini 2020). The different definitions focus on the lingual dynamics (atypical, deviant, deviated, or reverse swallowing) or consider swallowing in the broader context of often concomitant dysfunctions of the bucco-facial system (dysfunctional swallowing).

Atypical swallowing has a multifactorial etiology and is sometimes linked to dysmorphisms of the hard palate or airways, to adenoid hypertrophy or hypertrophic tonsils, to allergic rhinitis, or to postural anomalies of the skull, mandible, or tongue (Maspero et al. 2014).

© The Author(s), under exclusive license to Springer Nature
Switzerland AG 2025
A. Amitrano, F. Benso, *Atypical Deglutition Treatment*,
https://doi.org/10.1007/978-3-032-01482-5_3

Etiological factors of atypical swallowing are:

- Artificial feeding.
- Feeding with little variability of food consistencies.
- Oral vices.
- Oral anatomical alterations (altered lingual frenulum, macroglossia).
- Inflammatory diseases of the upper respiratory tract.
- General diseases (genetic syndromes).

The failure to transition from infantile to adult swallowing is a problem with a very high incidence in the population, in which only 85–90% of adults have correct swallowing (Proffit 2001) and about 60% of children between 4 and 8 years old have myofunctional alterations (Stahl et al. 2007). Rix (1946), evaluating a sample of 93 children aged between 7 and 12 years, found 61 with atypical swallowing. Werlich (1962) evaluated 640 elementary and middle school children and found that 30.4% had atypical swallowing. Rogers (1961), comparing a group of pediatric orthodontic patients with a sample of public school children, some of whom had orthodontic problems, noted that the incidence of atypical swallowing was high in both groups: 56.9% in schoolchildren and 62.8% in orthodontic patients. Atypical swallowing, due to its high incidence in the population and its correlation with den-toskeletal, postural, and functional problems more or less localized in the facial district, is a topic of significant interest in orthodontic, myological, and speech therapy fields (Maspero et al. 2014).

Atypical swallowing, while ensuring the transport of the bolus toward the gastric cavity, is carried out in a different way than the typical one for the subject's age. The differences concern the oral phase of swallowing, that is, the set of procedures that take place within the oral cavity, and concern the treatment of the bolus and its subsequent transport to the pharynx. It is useful to remember that the oral phase of swallowing, although responding to automatisms, is a voluntary phase that presupposes the activation of the cortical structures. The degrees of voluntariness vary as the swallowing act takes place. Voluntariness is broad and strongly influenced by motivational processes in welcoming the bolus when opening the mouth: Think of the child, but not only, who refuses a certain food by clenching his lips. However, even this phase is not free from automatisms that determine the degree of mouth opening. Rossi et al. (2020) have demonstrated how the degrees of mouth opening vary according to the size of the bolus to be introduced, well before the contact of the same with the lips. This highlights the existence of a predicting activity that allows the degree of opening to be programmed in a manner appropriate to the volume of the bolus to be welcomed and according to the instrument with which it reaches the lips, for example, the spoon rather than the straw. Voluntary control, therefore, is not exercised on the execution of individual movements of the oral structures but on the intention to start swallowing. In other words, when a bolus arrives in the pharynx or esophagus, the transport mechanisms are activated by reflex; vice versa, when a bolus arrives in the mouth, the bolus processing and transport mechanisms are activated only when the subject voluntarily decides and wishes to swallow (Groher 2021) (see CAP. 2).

First of all, we must get rid of the idea of a clear dichotomy between automatic and controlled processes: Cohen et al. (1990) try to make people understand that there is nothing absolute, and there is always a gradual interaction between the controlled systems that imply voluntariness and the automatic ones that manifest spontaneously, elicited by particular trigger stimuli that provoke the triggering. The authors argue, therefore, that the distinction between automatic and controlled processes is often discussed as if automation were of the "all or nothing" type. However, it seems quite clear that automation is rather configurable as a matter of gradation: The three scholars in fact express their strong suspicions that many discussions that treat automaticity as a dichotomous variable arise only because of the lack of a complete theoretical framework in which to formulate a more continuous relationship.

This further illustrates how some processes defined as "automatic," such as walking, assessed with dual tasks (see Sect. 2.8) reveal unequivocal degrees of interference (Patel et al. 2014; Rabaglietti et al. 2019). Also interesting is what was found in our recent experiment (Benso et al. (n.d.) in preparation; see Sect. 3.3) in which the interference between a very simple dual task and eating in subjects without any atypicality emerges. In other words, complex motor activities never completely discharge attentional control which, in the most automated forms, appears as a sort of light and unconscious monitoring. Given this premise, we must now make a distinction between the development of swallowing in a typical and atypical way. The development of swallowing in a typical situation occurs naturally in the early implicit stages: Once this ability is acquired in the adult state, a sort of idiosyncratic chewing rhythm (different from individual to individual) is established and likewise an optimal moment to swallow, also dependent on the consistency of the bolus. Therefore, when the voluntary act is not required by particular situations (e.g., in cases of unboned fish, large pills, abnormal mouthfuls, etc.), everything happens with a hyper-learned automatic process (in the sense of the term also defined by the aforementioned Cohen et al. 1990). Voluntariness can consciously emerge for some extemporaneous controls of the process, for example, when you want to change the rhythms for health purposes, resisting the interference of the acquired frequencies of the chewing and swallowing rhythm; for example, when in some yoga schools, you are invited to chew for a long time until the food "disappears from the mouth." In these specific cases, especially in the initial phases, the control of the CEN and the voluntary systems linked to attentional resources is very important.

In other words, from a certain point onward in typical development, the flow of food is followed with its own rhythm that sets the frequency and strength of chewing. The consistency of the food and/or the number of chews implicitly leads to swallowing from a certain phase of the process. This rhythm is hyper-learned, and if you watch TV while eating, you do not explicitly follow this moment of the process. You swallow without paying attention or without remembering having done so. This implies that voluntariness is not present at that moment but that the process in its frequencies of chewing and in the sensitivity to the consistency of the food finds that implicit evidence that acts as an idiosyncratic trigger to promote swallowing, a highly automated aspect and largely delegated to subcortical systems.

The situation is different for the subject who corrects swallowing because he will instead have to make extensive use of the voluntary control system. In this case, the automatic expertise achieved is not correct; therefore, the voluntary aspect emerges more consistently in the reeducation phases of a so-called "atypical" swallowing. In this case, several errors due to compensations also of physiological aspects or unfavorable dysmorphisms have become automated over time.

To improve the situation, in addition to the necessary physiological adjustment, programs must be set up in which it is important to work on the so-called "hyperlearning," in the sense of bringing a new and more correct process towards automation. In this case, as in all new and complex learning, the CEN and the executive control system are inserted and the processes practiced are largely dictated by the will. These processes, however, must move toward expertise, and this implies the gradual discharge of the CEN (see Sect. 2.6; Buccino et al. 2004; Vogt and Magnussen 2007; Benso et al. 2021; Benso, Benso, in preparation). This phase, however, is also very delicate because it must start the subject toward a consolidation of the gesture such that it can be established in any context: It is a critical moment, known to enablers when they notice that what is performed correctly in the clinical study is lost in everyday life.

The ABAED® method therefore intends to deconstruct nonadaptive automatisms, also rich in consolidated and disturbing errors, to bring back toward an initially voluntary swallowing that must be automated over time, cleaned of errors. However, it will then have to consolidate, overcoming the interference and the reemergence of previous spurious automatisms widely practiced over the years, and it will do so with methods that we will mention in Sect. 3.2.

Returning to the "historical" aspects inherent in swallowing theories, we continue with the models of Kennedy and Kent (1988), who theorized that swallowing involves the nervous system at three different levels: (1) peripheral, linked to the characteristics of the afferent bolus; (2) subcortical, which organizes and executes learned patterns of efferent activity; (3) cortical which responds to any necessary change in motor activity, based on perceived changes in the need to modify eating behavior. The lingual dynamics is very different in the infantile and adult swallowing, although the tongue does not exhaust the difference between one type of swallowing and the other (Table 3.1).

**Table 3.1** Swallowing patterns compared

| Atypical swallowing | Adult swallowing |
| --- | --- |
| Lip incompetence | Lip competence |
| Absence or presence of occlusal contact (eight Garliner's occlusal patterns) | Mild occlusal contact |
| Abnormal tongue support | Point of the tongue in contact with the retroincisal papilla or spot (Garliner 1976) |
| Positive intraoral pressure | Negative intraoral pressure |
| Contraction of the orbicularis oris, buccinator, and mentalis muscles | Absence of contraction of the perioral muscle |

The diversity is mainly related to the way the bolus is moved, which, in a typical swallow, occurs through progressive opposition of the tongue to the hard palate, creating differences in pressure gradients suitable for the progression of the bolus. The tip of the tongue moves upward toward the palate, posterior to the upper incisors, while the central part lowers, creating a central sliding channel, thanks to the lifting of the edges. Subsequently, the central part in turn rises toward the hard palate and so on, to create an increase in pressure upstream of the bolus. The oppositional movement of the tongue generates a pressure calculated between 680 and 2700 g (Garliner 1996), which causes occlusal problems if improperly discharged on the dental arches, as the tongue acts both as an impeding force and as a moving force (Garliner 1976; Ferrante 1995). Garliner distinguishes eight different types of occlusal patterns corresponding to as many incorrect positioning of the tongue during swallowing and consequent alterations of the oral muscular balances (Table 3.2).

The importance of correcting alterations in lingual dynamics and related oral dysfunctions to promote the correct evolution of dental bite, the physiological development of the cervical spine and the general postural balance of the child, is now an acquired knowledge in clinical and rehabilitative practice. In the broad scenario of rehabilitative methods, various techniques can be found, mainly related to the myofunctional approach, which has produced effective and widely used operational protocols. The current availability of new interpretative models of movement makes it possible to change perspective in the treatment of oral motor imbalances related to atypical swallowing and, more generally, of the motor aspects of swallowing. In fact, from a biomechanical and kinesiological vision of movement, the daughter of the correlation is the medicine of the nineteenth century, which with Bain in 1855 affirmed that the brain knows the muscles. We moved in 1882 to Jackson, according to whom the brain knows movements and not individual muscles. Subsequently, the Soviet neurophysiological school, with Anokhin, Lurija, and Bernsteijn, demonstrated that the brain knows the actions, and finally, more recently, the Rizzolatti group affirmed that the brain knows the goals. Furthermore, motor equivalence expresses the ability to achieve the same goal with different motor strategies and using different body districts: You can write the word "love" with your finger, your hand, and even with your foot (Berthoz 2019). As Berthoz (1997) states, the same separation between afference and efference no longer constitutes such a rigid barrier: "We must abandon the distinction between sensorial and motor; the boundaries between perception and movement are erased." In other words, sensations and perceptions on one side and muscles movement, action, and purpose on the other are to be considered two sides of the same coin. The Rizzolatti-Sinigaglia group (2006) states, "The same rigid boundary between perceptive, cognitive, and motor processes up revealing itself in largely artificial; not only does perception appear immersed in the dynamics of action... but the brain that acts is first and foremost a brain that understands. This understanding is pragmatic, preconceptual and prelinguistic and yet no less important since many of our much-celebrated cognitive abilities rest on it."

**Table 3.2** Occlusal patterns during atypical swallowing according to Garliner

| Type of atypical swallowing | Action of the tongue during swallowing | Orofacial musculature |
|---|---|---|
| 1a | The tongue pushes against the palatal surface of the upper incisors | The strength of the masseter, orbicularis, and mentalis muscles is normal |
| 1b | The tongue is inserted between the upper and lower incisors | Hypotonia of the orofacial muscles |
| 2 | The tongue is interposed between the dental arches from molar to molar | The masseter and the orbicularis muscles are hypotonic, and the mental muscle is overdeveloped |
| 3 | The tongue is interposed between the dental arches at the anterior level | The masseter muscle is in normal tone, the orbicularis muscle is weak, and the mental muscle is overdeveloped |
| 4 | The tongue pushes against the palatal surfaces of upper and lower incisors | The masseter muscle is tonic when there is posterior occlusal contact, the orbicularis muscle is hypotonic, and the mental muscle is hypercontracted |
| 5 | The tongue in a low position pushes mainly on the lingual surface of the lower incisors | The masseter and mental muscles are normotonic, and the orbicularis is hypertonic |
| 6 | The tongue interposes in the overjet of 1–2 mm which in this case is not on a genetic basis (genetic closed bite is not accompanied by atypical swallowing, AD, and muscle dysfunctions) | The masseter and orbicularis muscles are hypotonic, and the mental muscle is in normal tone |
| 7 | The tongue pushes unilaterally between the lateral incisor and the first molar | The orbicularis and mental muscles are normotonic, and the masseter muscle is hypotonic on the lingual thrust side |
| 8 | The tongue pushes bilaterally between the lateral incisor and the first molar | The orbicularis and mental muscles are in normal tone, and the masseter muscle is hypotonic on both sides of the lingual thrust |

*Source*: Adapted from Garliner (1976)

## 3.2   The ABAED® Method

The ABAED® method is a (re)habilitation methodology aimed at promoting the transition from infantile swallowing to adult swallowing. The working hypothesis is the following: The failure to transition from infant swallowing to adult swallowing constitutes the failure to learn a motor ability (skill), at least in cases where the anatomical alterations of the oral structures do not hinder such transition. Swallowing in fact cannot be considered a simple connection between the outside and the inside, allowing the introduction of food substances into the digestive system, but is a set of movements of various kinds that, in an effective and safe manner, convey food from the lips to the stomach. The term "movement" defines the change of even minimal spatial positions, of the body and of its individual parts (Schmidt 1982). "Movement," "motor act," and "action" are the terms commonly used to describe movement and, although often used as synonyms, in reality they have different meanings for scholars of the neuronal substrates of motor behavior (Aglioti and Facchini 2002; Rizzolatti and Sinigaglia 2006; Mandolesi 2012; Table 3.3).

Motor acts and actions represent the highest degree of complexity, are based on intentionality, and are planned according to a purpose. Such properties do not exhaust the set of characteristics of movement. Motor acts and actions are defined as voluntary. They respond to the activity of certain cortical areas and are generally learned and can be improved with exercise. Some motor behaviors are voluntary only at the beginning and at the end and are therefore defined as "rhythmic"; this group includes behaviors such as walking, chewing, and running. A third category of motor behaviors is made up of reflexes, that is, rapid and stereotyped responses to a specific stimulus (Ghez 2003). Motor behaviors can also be classified based on the control mechanisms and sensory information that allow for any corrections (Aglioti and Facchini 2002). In this sense, we can distinguish between closed-circuit motor behaviors and open-circuit motor behaviors. In the realization of closed-loop, feedback, or adaptive behaviors (the feedback closes the circuit), sensory information constantly monitors the execution of the behaviors themselves; for example, keeping an inflated balloon in the air is included in these behaviors, as the action is directed by sensory information that comes from the constant analysis of the balloon's trajectory. Closed-loop motor behaviors are, due to the ways in which they are realized, slow in execution. We could say, with the terminology outlined in Chap. 2 on attentive networks, that the entry of the CEN control system and working memory slows down the processes that are thus monitored almost "online."

Open-loop motor behaviors, or feedforward, are not carried out based on the constant analysis of sensations as they are very fast and are essentially based on prediction mechanisms (the expertise discharges the CEN). This category includes, for example, a goalkeeper's save during a penalty kick: The speed of the ball and the short distance of the shooter prevent a constant analysis of the sensory information and therefore the motor behaviors of the person defending the goal are based on a prediction. The goalkeeper, in fact, decides a priori where to dive and his behavior will be effective or not depending on a random prediction, which does not use a step-by-step analysis of the action. The same motor behaviors can be carried out in

**Table 3.3** Motor behaviors

| | |
|---|---|
| *Movement* | Activation of a small muscle district that determines the movement in space of one or more joints |
| *Motor act* | Multiple movements performed in a synergistic and fluid manner to achieve a goal |
| *Action* | Sequence of motor acts aimed at achieving a goal |

open or closed mode in different situations; for example, walking normally occurs using an open-circuit behavior, but if we are busy walking on a wet floor, the same behavior is carried out with a closed-circuit scheme. Due to the different characteristics of a motor behavior, when designing an enabling or rehabilitative intervention, it is useful to first define the type of motor behavior that is the object of the intervention.

The oral phase of swallowing includes motor behaviors that can be classified as motor acts and actions, that is, voluntary behaviors based on intentionality, which have a purpose and are planned. Oral motor behaviors are rhythmic in nature and, depending on the type of food and the type of situations, are carried out using an open- or closed-circuit scheme: Think, for example, when we eat a piece of bread and when instead we eat a piece of fish that may contain bones. These motor behaviors can be improved with exercise and, in most cases, are learned movements. The performance of motor behaviors "is always the visible result of a series of neuronal processes that involve the activation of many brain circuits that are not only in the motor domain" (Mandolesi 2012). When we act, we know what to achieve and most of the times how to do it; we can assume that the atypical swallower knows what to achieve, that is, to bring food to the stomach, but does not know how to do it or, better, his know-how is not typical of his age. When we are in the condition of not knowing how to do, we resort to learning mechanisms or, to put it in the words of the 2000 Nobel Prize winner for Medicine, Eric Kandel, "learning is the biological basis of individuality: through learning man forms his own being, his own personality" (Kandel and Schawartz 1991). The ABAED® method aims to facilitate the motor learning processes of adult swallowing in children and adults. In recent years, Amitrano and Benso have studied an intervention protocol on atypical swallowing, based on the most recent acquisitions related to motor learning (motor learning). The ABAED® approach overturns the perspective of the intervention on atypical swallowing, which is no longer addressed to the orofacial muscles but to the brain, the organ that understands, decides, plans, implements, and monitors the execution of motor programs. In other words, the intervention is carried out with an up-down strategy (from top to bottom, i.e., from the brain to the muscle) rather than with a bottom-up strategy (from bottom to top, i.e., from the muscle to the brain). The method also includes techniques that are carried out with conscious maneuvers and that require loads of executive attentive resources and working memory. In the field of neuroscience, it is argued that the strengthening and recall of executive attention during learning processes facilitate and speed up the acquisition of the skill being treated (Wagner et al. 1998; Blumenfeld et al. 2006; Seidler et al. 2012; Benso 2018; Benso et al. 2021). Furthermore, proprioceptive and somatognosia awareness, techniques based on mental imagery and process visualization, and the maintenance of a "here and now" awareness during learning phases require an attentional system focused and centered, as well as a working memory with adequate resources. Motor learning or skill acquisition "is an internal state or process, whose level reflects a person's ability to produce a movement at a given moment" (Schmidt and Wrisberg 2008). The learning of a skill can be implicit or explicit. Implicit learning is a learning modality that occurs without awareness, is innate, and is based on experience. In the first years of life, the implicit mode is the child's only form of learning. Explicit learning is based on instructions and explanations and presupposes conscious adherence to a provided model. The physiological transition from the infantile swallowing mode to the adult one occurs through implicit learning, while in atypical swallowers, the transition occurs through a formal path and therefore through explicit learning. There are three variables that influence explicit learning: the person, the task, and the environment. As for the person, the level of maturation, motivation to learn, cultural level, and emotional participation must be considered. The task instead requires a task analysis and verification of the anatomical-physiological conditions necessary for the execution of the task that is to be learned. Lastly, an analysis of the environment is necessary, that is, the conditions in which the learned behavior must be applied. Before undertaking the ABAED® program, it is necessary to verify the anatomical-functional adequacy of the organs involved in the development of adult-type swallowing. The presence of a short lingual frenulum, an ogival palate, or the presence of orthodontic appliances occupying the oral cavity and in particular the palatal vault should be considered as a contraindication to learning the adult swallowing mode. It is in fact essential that, during the learning process, the tongue can perform its movements freely in the adequate spaces. The analysis of the subject's cultural level is important in choosing the most suitable language so that all the information necessary for the application of the method is well understood. An important prerequisite for effective learning is the clear understanding by the student of the learning objective (Schmidt and Wrisberg 2008). During the first ABAED® treatment session, the objectives are explained using methods and terminology suitable for the subject's age. The explanation is repeated at the beginning of each session, until the understanding of the objective appears fully consolidated. It is important to verify the patient's motivation and evaluate the need for a motivational path while it is worth the opportunity not to undertake ABAED® in case of absent motivation or evident opposition. Learning a new motor skill of swallowing requires a period of practice aimed at achieving the ability to perform the new task in an automatic, economical, fast, and precise manner (Mandolesi 2012). The acquisition of a motor skill involves three phases: cognitive, associative, and automation (Fitts 1964; Prosiegel et al. 2000; Buccino et al. 2004; Abernethy et al. 2005). During the cognitive phase, the subject learning the motor skill must understand what to do and what are the purposes of the action to be acquired and used, in order to apply the rules that he begins to translate into motor acts. The subject also uses subvocal verbalization to repeat the movements he is performing and produces poor performances, with many errors and uneconomical movements, based on co-contraction motor strategies. Co-contraction is a costly strategy of simultaneous contraction of the flexors and extensors, which the CNS implements whenever it does not know the load (or this is too heavy, as for example happens in weightlifting) or the movement is not known. The co-contraction mechanism is abandoned when the movement is learned or the load known, in favor of the more economical mechanism of reciprocal innervation which sees the activation of the flexor and simultaneously the inhibition of the extensor or vice versa. During the associative phase, the subject learning a motor skill associates individual movements in structured and purposeful motor sequences and uses mental representation to better control the action, while attention is mainly directed at salient points. Repetitive exercise allows the transition to the next phase of automation. In cases where the task is performed with errors, it is necessary to avoid it being memorized and return to the cognitive phase, because the correction of an automated sequence is much more difficult. During the automation phase, the motor skill is gradually automated and requires a progressively lower attention load. The times for automation are long and, for more complex actions, it is probably a task that requires very long times. The estimated times needed for the complete automation of a motor skill would be about 10 years or 10,000 h of training (Abernethy et al. 2005). All this comes from a study by Ericsson et al. (1993). In 2014, a book by Malcolm Gladwell (Outliers. The Story of Success) was published in Italy, which refers to that theory. The challenge was rather to ask how it was possible to shorten the time to become experts. We found some strong clues in the past from research in the so-called "flow states" introduced in the 70s by psychologist Mihály Csíkszentmihályi, of Claremont Graduate University, in California. He had noticed a particular unforced state of attention in children who learn faster than others. Csikszentmihályi (1997) compares this state (defined as "flow") to the changes of critical thinking that are obtained during periods of intense concentration, such as Zen sessions, during which the most advanced practitioners discharge ruminating thoughts and enter a "here and now" awareness. According to Csíkszentmihályi, calming self-critical thoughts could free those more spontaneous processes, which in turn would produce that concentrated sensation of effortless "flow," during which one learns faster what is the fruit of our current application.

In ABAED®, especially in the pretreatment phase, we take into account these aspects at least in part (Benso, Benso, in preparation). Furthermore, to reduce the time of the automation phase, as explained in Chap. 2, we evaluate the usefulness of keeping some brain circuits implicated in awareness and learning (Wagner et al. 1998; Blumenfeld et al. 2006), and we do this by pre-activating particular attention states and using the dual-task technique (calibrated and adapted to the subject and the context) which as we know (see Sect. 2.8) from Varela-Vásquez et al. (2020) speeds up learning of the main task.

The more technical "speech therapy" part of the ABAED® method includes two distinct and consecutive treatment phases: In the first, the subject learns the lingual dynamics typical of adult-type swallowing. In the second, more "attentive speech therapy," the learnings are consolidated and automated. In the first phase, the subject, using mental images, proprioceptive experiences, executive attention, and working memory, learns the lingual movements of adult swallowing. In the second phase, through the use of dual tasks, the subject automates the learnings.

Each session follows a precise four-phase structure in which exercises specific to the treatment phase are inserted:

1. Attentive activation and centering of arousal.
2. Application of treatment.

3. Facilitation of learning and consolidation of the motor process, using dual tasks tailored to the subject.
4. Enhancement of working memory with the function of facilitating motor learning.

As already stated, an important prerequisite for effective learning is a clear understanding, by the learner, of the learning objective. Therefore, in the first session and in the following ones, until necessary, the treatment objectives are explained. Then the subject is helped to perceive and recognize the parts of the mouth and specifically some points of the tongue and palate that correspond to the anterior, central, and posterior parts. The subject learns to feel and identify the points indicated in their own mouth and in an external model (a drawing or his own hand). Subsequently, the subject observes the lingual dynamics of adult-type swallowing. Since the tongue movements are realized inside the closed mouth, the observation is carried out with visualization techniques that, from time to time, use films, drawings, or mental images (Jeannerod 1994, 1995; Table 3.4).

In ABAED®, internal and external mental images are used to visualize the individual anatomical parts involved in swallowing and the movement of the tongue in its entirety. In this phase, it is important that the movements are not only observed but also understood in their purpose and in their dynamics of execution. To facilitate the understanding and realization of the new lingual dynamics in the swallowing phase, the articulatory movements previously made conscious are used. In other words, the subject is made aware of some articulatory movements that he habitually produces and subsequently learns to reuse the same movements in swallowing function, first through simulations made with movements of the hand and then with the tongue.

The readaptation of preexisting abilities accelerates the process of learning new motor skills (D'Avella 2016). The execution of the new lingual dynamics is constantly monitored with mental images, whose verbalization allows the speech therapist to control the correctness of the lingual movements. Subsequently, the new lingual dynamics learned is used in swallowing tasks of foods of different consistency. Swallowing of boluses is performed according to a precise stratification of complexity. We start by swallowing creamy boluses, then solid boluses, and only lastly liquid boluses which, due to their rheological characteristics, constitute the highest level of difficulty. This part of the treatment requires a specific setting. In fact, the exercises are performed in front of the mirror in order to provide the subject with visual feedback that facilitates the correct execution of the task. The first exercises involve swallowing, with adult-type lingual dynamics, of small creamy boluses that the speech therapist initially places on the tongue; subsequently, the patient

**Table 3.4** Types of mental images

| Internal | When we mentally simulate the execution of a certain action, whether it concerns the whole body or a part of it |
|---|---|
| External | When we imagine a certain scene or an object |

**Table 3.5**  Types of feedback

| Intrinsic (internal) | Feedback that is sensorially perceived during the execution of the task |
|---|---|
| Extrinsic (external or augmented) | Feedback that provides information on performance that supplements the information coming from intrinsic feedback and is generally provided by an external agent (teacher, coach, therapist, etc.) and is provided on the performance or on the result |

independently places boluses of increasing size in his or her mouth before swallowing them using the correct mode. In this phase, the patient uses proprioceptive and visual feedback and has an extrinsic feedback on performance and result, provided by the speech therapist (Anderson et al. 2020; Table 3.5).

When all about 40 boluses planned for training in a single session are consistently swallowed correctly, that is, without the presence of abnormalities in tongue movements, we move on to the next phase, that is, training with solid boluses (non-crumbly biscuits or other). The structuring of the setting and the feedback methods remain unchanged compared to training with creamy boluses. Also in this case, the correct consecutive intake of 40 solid boluses indicates the possibility of moving on to the next step, which is the swallowing of liquid boluses. Small boluses of water are drunk with the mouth closed, always in front of the mirror. In this case, the patient has proprioceptive feedback and the external feedback from the speech therapist. When even the swallowing of 40 liquid boluses does not differ from the adult swallowing mode, the first part of the treatment can be considered concluded. All the foods used in this phase are chosen respecting the progression of consistencies indicated by the method, considering the preferences and any allergies or food intolerances of the subject. It is important that the foods used respond to the tastes of the subject for the impact that the choice can have on the motivation to treatment, even if some choices are sometimes contrary to the indications of orthodontists, who discourage the use of sweet substances.

It is important to underline that each ABAED® session, regardless of the treatment phase, begins with the typical pretreatment typical of Integrated Cognitive Training (ICT; see Benso 2018; Benso et al. 2021; Benso, Benso, in preparation). To better understand the spirit of the maneuvers that precede and that are an unavoidable backdrop of the ABAED® method, let's reflect on an aspect that risks being overlooked due to its glaring evidence. After a day of school, the child, who goes to a clinical center, could be tired or too excited by existential stress; he may have depressive thoughts while going to lock himself in a clinical studio while his friends have fun in the gardens with their bicycles. In other words, the subject may present for a diagnosis or treatment tired, demotivated, grumpy, and attentively not centered. In this case, the knowledge of executive attention models (see Chap. 2) is useful for a concrete application to clinical reality. We note that Malinowski (2013), describing mindfulness techniques, cites in the processes the same identical attentional networks. In the pretreatment, a welcoming relationship should be foreseen, which stimulates the subject with appropriate psychophysical and attentive activation techniques, through appropriate exercises, starting from phasic alert. In the treatment phase, as mentioned, to improve swallowing, specific speech therapy exercises are administered that already include work in awareness and, at specific moments, visualizations of the motor image (hence the usefulness of exercises also performed "apart" that strengthen this ability). In the second phase, the enabling activity will use and continue introducing "double tasks." In ABAED®, double tasks are used above all in the consolidation phases. This phase includes feeding exercises with predominantly solid boluses that are swallowed in double-task situations: The subject is in fact asked to simultaneously perform two tasks, one of which consists of eating. This technique will have various functions, among which that of strengthening working memory, of favoring learning times, and of promoting the assembly of the various movements still broken down into a single motor act, to keep active the functional circuits for memorization, which would be more easily deactivated with automated practices carried out absent-mindedly (see D'Esposito et al. 2000; Wagner et al. 1998). Such stereotyped practices would fall into a "nonadaptive" and unproductive situation, as demonstrated by Metzler-Baddeley et al. (2016, 2018). Subsequently, always using the dual tasks with different calibrations, we will proceed to consolidation of the newly acquired motor act. Given the importance of the resources in working memory necessary to fuel these processes (Buccino et al. 2004), these will also be measured and stimulated separately, as well as motor imagination.

The importance of choosing the appropriate dual task for achieving the different useful purposes in which double tasks can be used (see Sect. 2.8) should be highlighted. The double task chosen is not always the optimal or useful one for that specific purpose, and in some cases, the choice of a double task not well calibrated to the subject can destroy the performance instead of promoting it.

Finally, it should be remembered that in ABAED®, implicit learning proposals for the correct lingual positioning at rest are not excluded, in the complicated consolidation phase, through mobile instruments to be kept especially during the night hours; this type of intervention is carried out in close collaboration with the orthodontist.

Once the formal learning process of adult swallowing is completed, the subject continues to use the learned lingual dynamics in contextual situations and follows a follow-up program. The treatment is generally completed in 20 sessions with a weekly interval. The patient independently, or with an informed caregiver, performs daily workouts in the ways indicated by the speech therapist, depending on the phases of the treatment. The total duration of the treatment—especially of the more definitive consolidation, to be accompanied with double tasks and monitoring in home environments, not just clinical—is to be considered indicative, because the passages from one step to another follow the learning curve of the subject.

## 3.3　Experimental Study

Some parts of ABAED® have been subjected to experimental studies; specifically to study the use of a dual task during the oral phase of swallowing, preparation, and transport of boluses, an experiment was set up from the need to study some types of dual tasks to adapt to the swallowing learning phases (Benso et al., in preparation).

Seemingly automated processes like walking are circuits open to top-down processes so that, when necessary, the control system can intervene to protect the subject's safety, notwithstanding, as already described, that an excess of emotional activation can engage or disengage the CEN systems even in a nonadaptive way. As already mentioned, it is important to consider that the use of the dual task can improve the times and quality of learning including the enabling of the primary task. Varela-Vásquez et al. (2020), in a review of many works mainly aimed at improving balance and gait speed in the elderly to reduce the risk of falls, conclude by saying that it is relevant to note that all studies show clear improvements in all participants, both among those who performed training activities in a single task and among those who performed training programs in dual task. However, the results of training in dual task are superior to training in a single task. As expressed in Sect. 2.8, we reiterate that a dual task can be used to assess the degree of automaticity of the main task. If no differences are found in the concurrent administration compared to when the tasks are performed individually, this means that at least one of the two tasks is so closed and automated that no mutual interference is created. A result of low interference with a system that has become automated indicates little possibility of destructuring and therefore of improvement. We decided to investigate the process of walking in literature, for the possibilities of expressing very automated or very controlled states, to then move directly to evaluate chewing and swallowing with the same intent. On walking, there are conflicting results: Some authors do not find interference in the performance of walking, while others, like Rabaglietti et al. (2019), with probably more complex tasks find that, regardless of age, a dual-task performance could influence walking performance based on the secondary task required. Furthermore, their results have shown the association between working memory skills and the cost of the dual task in walking ability. Several other works in this sense find interference and we can mention among all Patel et al. (2014).

As far as we are concerned, it is not so much the thought of having to interfere on swallowing itself to demonstrate that the system is as open as walking, but rather we consider the reverse aspect, in our opinion more sensitive and, above all, easier to measure. In the sense, it is good to assess whether and how much chewing and swallowing subtract resources to a dual task that requires sustained attention for 5 min. In other words, if the "chewing and swallowing" system were completely closed and totally automated, it would not interfere with a simple reaction time task; on the contrary, it would require control vigilance, at least occasional, which can interfere with a concomitant task of sustained attention. The question has a certain importance because if we face a "modularly closed" and therefore super-automated system, there are few possibilities to deconstruct and intervene with treatments or other. Therefore, the aim of this study was to try to understand if the "chew and swallow" system could or could not influence an additional task that required a modest commitment of sustained attention. The hypothesis is the following: If the process of chewing and swallowing is a system that, although automated, requires occasional attentive checks such as walking, then a concomitant task of sustained attention to reaction times should reveal slowdowns. This would indicate the possibility of being able to act on swallowing processes that have developed anomalously, which however can still be deconstructed to reconstruct a typicality of the movement.

In detail, 24 students from the third grade of a middle school, with the authorization of the school board and the consent of their parents, had to respond as quickly as possible, pressing a button (reaction time or Reaction Time, RT) to a pressing beep in a single-task condition and in a dual-task condition (simultaneously eating crackers). A computerized program emitted 90 beeps for 5 min and 9 s. The Inter-Stimulus Interval (ISI) was represented by three different intervals: 3000 ms, 3500 ms, and 3800 ms. Each of the three groups of stimuli composed of 30 beeps appeared randomly. The subject had to respond each time by pressing a button and the RT parameter was recorded by the computer. The subjects in the single-button task of response to the succession of beeps showed an average RT of 411.66 ms, with a standard deviation of 42.71 ms; when instead they had to eat, the times rose significantly; the average RT was 558.02 ms and a deviation of 173.04 ms. The Mann-Whitney U test was significant as the difference between a single task or dual tasks, with a $P < 0.001$. The increase in reaction times in the dual-task phase indicates that chewing and swallowing are not zero-attention-cost activities, but a sort of vigilance may be required as implicit control. Such control may cause the sequence of beeps to be lost, and the increase in deviation found probably provides evidence of the redirection of attention toward swallowing the cracker.

This result indicates the possibility of building intervention methods that also rely on the fertile push for improvement that dual tasks provide. The ABAED® method, also based on this evidence and others we have tested, uses dual tasks especially in the consolidation phase of the treatment.

The application of the ABAED® method, like all trainings that aim to increase the efficiency of the results, also relies on the conscious preparation of the operators. Such preparation requires expert knowledge of the proposed speech therapy technique, which must be carefully studied and assimilated, even if knowledge alone is not sufficient to address the problems related to learning that then have difficulty in establishing themselves and above all in consolidating themselves in everyday life, as is well known to those in the field. Therefore, the combination and above all the knowledge of the attention networks, of the methods of manipulating their states, even in pretreatment, of the neuroscientific models of working memory and their precious support for learning in general and specifically for motor learning, become essential. This also implies the use of techniques to enhance working memory (re-elaborations, task shifting, n back), exercises in imagination, and memory of gestures and the expert use of double tasks: all made functional to the proposed speech therapy exercises. One aspect not to be forgotten, to make the treatment functional and efficient, is that the sequence of techniques to be administered systematically is not the only point to be privileged, but of absolute relevance is the figure of the speech therapist with his neuroscientific knowledge of models and functions and his ability to the attentive functions in subjects also from an emotional and relational point of view.

# Suggested Reading


Abernethy B, et al. The biophysical foundations of human movement. South Yarra, VIC: Macmillan Education; 2005.

Aglioti SM, Facchini S. Il cervello motorio. In: Spinelli D, editor. Psicologia dello sport e del movimento umano. Bologna: Zanichelli; 2002.

Agostoni C, et al. Complementary feeding: a commentary by the ESPGHAN committee on nutrition. J Pediatr Gastroenterol Nutr. 2008;46(1):99–110.

Amitrano A. Dyspagia and nutrition. Springer; 2024.

Anderson DI, et al. Enhancing motor skill acquisition with augmented feedback. In: Hodges NJ, Williams AM, editors. Skill acquisition in sport. 3th ed. London: Routledge; 2020.

Andretta P. La deglutizione deviata e lo squilibrio muscolare orofacciale. In: Ruoppolo G, et al., editors. Manuale di foniatria e logopedia. Roma: SEU; 2012.

Andrew BL. The nervous control of the cervical oesophagus of the rat during the swallowing. J Physiol. 1956;134:729–40.

Andrews-Hanna JR. The Brain's default network and its adaptive role in internal mentation. Neurosci. 2012; https://doi.org/10.1177/1073858411403316.

Atkinson JC, et al. Major salivary gland function in primary Sjögren's syndrome and its relationship to clinical features. J Rheumatol. 1990;17(3):318–22.

Baddeley AD, Hitch GJ. Working memory. In: Bower GH, editor. The psychology of learning and motivation, vol. 8. New York: Academic Press; 1974. p. 47–89.

Benso F. Neuropsicologia dell'attenzione. In: Teoria e trattamenti nei disturbi di apprendimento. Pisa: Edizioni Del Cerro; 2004.

Benso F, Benso E. I principi neuroscientifici dei training cognitivi. Firenze: Hogrefe Editore; 2023.

Benso F, et al. The time course of attentional focusing. Eur J Cogn Psychol. 1998;10:373–88.

Benso F, et al. Principles of integrated cognitive training for executive attention: application to an instrumental skill. Front Psycol. 2021;12:647749.

Benso F, et al. (in preparazione), titolo da definire; n.d.

Bernstein JH, Waber DR. Executive capacities from a developmental perspective. In: Meltzer L, editor. Executive function in education from theory to practice. New York: The Guilford Press; 2007. p. 39–54.

Berthoz A. Le sens du mouvement. Paris: Odiel Jacob; 1997.

Blumenfeld PC, Kempler TM, Krajcik JS. Motivation and cognitive engagement in learning environments. In: Sawyer RK, editor. The Cambridge handbook of the learning sciences. Cambridge: Cambridge University Press; 2006. p. 475–88.

Bravo MJ, Nakajama K. The role of attention in different visual-search tasks. Percept Psychophys. 1992;51:465–72.

Buccino G, Binkofski F, Riggio L. The mirror neuron system and action recognition. Brain Lang. 2004;89(2):370–6.

Buettner A, et al. Observation of the swallowing process by application of videofluoroscopy and real-time magnetic resonance imaging-consequences for retronasal aroma stimulation. Chem Senses. 2001;26(9):1211–9.

Butler SG, et al. Factors influencing aspiration during swallowing in healthy older adults. Laryngoscope. 2010;120:2147–52.

Butler SG, et al. The relationship of aspiration status with tongue and handgrip strength in healthy older adults. J Gerontol A Biol Sci Med Sci. 2011;66:452–8.

Cain WS, Stevens JC. Uniformity of olfactory loss in aging. Ann N Y Acad Sci. 1989;561:29–38.

Calhoun KH, et al. Age-related changes in oral sensation. Laryngoscope. 1992;102(2):109–16.

Carlsson GE. Masticatory efficiency: the effect of age, the lost of teeth and prosthetic rehabilitation. Int Dent J. 1984;34:93–7.

Castiello U, Umiltà C. Temporal dimensions of mental effort in different sports. Int J Sport Psychol. 1988;19(3):199–210.

Ciarmiello A, et al. FDG-PET in the evaluation of brain metabolic changes induced by cognitive stimulation in mci subjects. Curr Radiopharm. 2015;8(1):69–75.

Clavé P, Verdaguer A, Arreola V. Oral-pharyngeal dysphagia in the elderly. Med Clin. 2005;124:742–8.

Cohen JD, Dunbar K, McClelland JL. On the control of automatic processes: a parallel distributed processing account of the stroop effect. Psychol Rev. 1990;97:332–61.

Comfort A. Ageing: the biology of senescence. London: Routledge & Kegan Paul; 1964.

Cook IJ, et al. Opening mechanisms of the human upper esophageal sphincter. Am J Physiol. 1989;257(5 (Pt. 1)):G748–59.

Cook IJ, et al. Influence of aging on oral-pharyngeal bolus transit and clearance during swallowing: scintigraphic study. Am J Physiol. 1994;266(6 (Pt. 1)):G972–7.

Coppo E, Belforte G, Corbetta L. La tuba di Eustachio. Ann Laringol Otol Rinol Faringol. 1966;65(1 suppl.)

Corbetta M, Shulman GL. Control of goal-directed and stimulus-driven attention in the brain. Nat Rev Neurosci. 2002;3:201–15.

Costa A, et al. Standardization and normative data obtained in the Italian population for a new verba fluency instrument, the phonemic/semantic alternate fluency test. Neurol Sci. 2014;35(3):365–72.

Cowan N. Evolving conceptions of memory storage, selective attention, and their mutual constraints within the human information-processing system. Psychol Bull. 1988;104(2):163–91.

Cowan N, et al. Deconfounding serial recall. J Mem Lang. 2002;46(1):153–77.

Cowan N, et al. On the capacity of attention: its estimation and its role in working memory and cognitive aptitudes. Cogn Psychol. 2005;51(1):42–100.

Cozza P, Loberto S, Andretta P. Deglutizione. In: Levrini L, editor. Terapia miofunzionale orofacciale. Milano: EDRA; 2020.

Csikszentmihályi M. Creativity: flow and the Psychology of Discovery and Invention. New York: HarperCollins; 1997.

D'Ambrosio M, Bracco F, Benso F. Misurazioni di fluenza in persone con balbuzie in doppi compiti complessi automatizzati e non automatizzati. Sistemi Intell. 2016;28(1):83–104.

D'Avella A. Modularity for motor control and motor learning. Adv Exp Med Biol. 2016;957:3–19.

D'Esposito M, Postle BR. The cognitive neuroscience of working memory. Annu Rev Psychol. 2015;66:115–42.

D'Esposito M, Postle BR, Rypma B. Prefrontal cortical contributions to working memory: evidence from event-related fMRI studies. Exp Brain Res. 2000;133(1):3–11.

Daniels SK, et al. Mechanism of sequential swallowing during straw drinking in healthy Young and older adults. J Speech Lang Hear Res. 2004;47(1):33–45.

Delaney A. Oral-Motor Movement Patterns in Feeding Development, Thesis for PhD, University of Wisconsin-Madison, Madison (WI); 2010.

Diamond A. Close interrelation of motor development and cognitive development and of the cerebellum and prefrontal cortex. Child Dev. 2000;71(1):44–56.

Dickstein SG, et al. The neural correlates of attention deficit hyperactivity disorder: an ALE meta-analysis. J Child Psychol Psychiatry. 2006;47(10):1051–62.

Dixon ML, et al. Interactions between the default network and dorsal attention network vary across default subsystems, time, and cognitive states. Neuroimage. 2017;147:632–49.

Dodds WJ. Physiology of swallowing. Dysphagia. 1989;3(4):171–8.

Dodds WJ, Stewart ET, Logemann JA. Physiology and radiology of the normal oral and pharyngeal phases of swallowing. Am J Roentgenol. 1990;154:953–63.

Dodrill P. Infant feeding development and dysphagia. J Gastroenterol Hepatol Res. 2014;3(5).

Dosenbach NU, et al. A dual-networks architecture of top-down control. Trends Cogn Sci. 2008;12(3):99–105.

Doty RW. Neural organization of deglutition. In: Code CF, editor. Handbook of physiology. Alimentary canal, sect. 6, vol. Iv. Washington DC: American Physiological Society; 1968.

Doty RL. Olfactory communication in humans. Chem Senses. 1981;6(4):351–76.

Ebihara S, Ebihara T, Kohzuki M. Effect of aging on cough and swallowing reflexes: implications for preventing aspiration pneumonia. Lung. 2012;190:29–33.

Eckert MA, et al. At the heart of the ventral attention system: the right anterior insula. Hum Brain Mapp. 2009;30(8):2530–41.

Einstein A. Pensieri di un uomo curioso. Milano: Mondadori; 1999.

Engen T. La percezione degli odori. Roma: Armando; 1989.

Engle RW, Kane MJ. The role of prefrontal cortex in working-memory capacity, executive attention, and general fluid intelligence: an individual-differences perspective. Psychon Bull Rev. 2002;9:637–71.

Engle RW, Kane MJ, Tuhoski SW. Individual differences working memory capacity and what they tell us about controlled attention, general fluid intelligence and functions of the prefrontal cortex. In: Miyake A, Shah P, editors. Models of mechanism of active maintenance and executive control. Cambridge: Cambridge University Press; 1999. p. 102–34.

Ericsson KA, Krampe RT, Tesch-Römer C. The role of deliberate practice in the acquisition of expert performance. Psychol Rev. 1993;100:363–406.

Ertekin C, Aydogdu I. Electromyography of human cricopharyngeal muscle of the upper esophageal sphincter. Muscle Nerve. 2002;26(6):729–39.

Etkin A, Wager TD. Functional neuroimaging of anxiety: a meta-analysis of emotional processing in PTSD, social anxiety disorder, and specific phobia. Am J Psychiatr. 2007;164(10):1476–88.

Fassbender C, et al. A lack of default network suppression is linked to increased distractibility in ADHD. Brain Res. 2009;1273(1):114–28.

Feldman RS, et al. Aging and mastication: changes in performance and in the swallowing threshold with natural dentition. J Am Geriatr Soc. 1980;28(3):97–103.

Ferrante A. La deglutizione atipica. Il Dent Moderno. 1995;15:227–39.

Fitts PM. Perceptual-motor skill learning. Categories Human Learn. 1964;47:381–91.

Fox MD, et al. The human brain is intrinsically organized into dynamic, anticorrelated functional networks. Proc Natl Acad Sci U S A. 2005;102(27):9673–8.

Franks HA, et al. Mechanism of intra-oral transport in a herbivore, the Hyrax (Procavia Syriacus). Arch Oral Biol. 1985;30(7):539–44.

Fried LP, et al. Untangling the concepts of disability, frailty, and comorbidity: implications for improved targeting and care. J Gerontol Ser A Biol Med Sci. 2004;59(3):255–63.

Fukunaga A, Uematsu H, Sugimoto K. Influences of aging on taste perception and oral somatic sensation. J Gerontol Ser A Biol Sci Med Sci. 2005;60(1):109–13.

Gandhi A. Vivi come se dovessi morire domani. v. Firenze: Giunti; 2019.

Gardner S, Merestein G. Handbook of neonatal intensive care. St. Louis (MO): Mosby; 2002.

Garliner D. Myofunctional therapy. Philadelphia, PA: W. B. Saunders; 1976.

Garliner D. Importanza di una deglutizione corretta, Futura publishing society; 1996.

German RZ, et al. Food transport through the anterior oral cavity in macaques. Am J Phys Anthropol. 1989;80:369–77.

Ghez C. Il controllo del movimento. In: Kandel ER, Scwartz JH, Jessel TM, editors. Principi di neuroscienze. 2nd ed. Milano: Casa Editrice Ambrosiana; 2003.

Gladwell M. Fuoriclasse: storia naturale del successo. Milano: Mondadori; 2014.

Good-Fratturelli MD, Curlee RF, Holle J. Prevalence and nature of dysphagia in VA patients with COPD referred for videofluoroscopic swallow examination. J Commun Disord. 2000;33(2):93–110.

Greicius MD, et al. Functional connectivity in the resting brain: a network analysis of the default mode hypothesis. Proc Natl Acad Sci U S A. 2003;100(1):253–8.

Greicius MD, et al. Resting-state functional connectivity reflects structural connectivity in the default mode network. Cereb Cortex. 2009;19(1):72–8.

Groher ME. Normal swallowing in adult. In: Groher ME, Crary MA, editors. Dysphagia clinical management in adult and children. New York: Elsevier; 2021.

Hamdy S. Cortical activation during human volitional swallowing: an event-related fmzu study. Am J Physiol. 1999;277(Pt. 1):219–25.

Hamdy S, et al. The cortical topography of human swallowing musculature in health and disease. Nat Med. 1996;2(11):1217–24.

Hanson ML, Cohen MS. Effects of form and function on swallowing and the developing dentition. Am J Orthod. 1973;64(1):63–82.

Hårdem Ark Cedborg AI, et al. Breathing and swallowing in normal man-effects of changes in body position, bolus types, and respiratory drive. Neurogastroenterol Motil. 2010;22(1L):1201–8. e316

Heath MR. The effect of maximum biting force and bone loss upon masticatory function and dietary selection of the elderly. Int Dent J. 1982;32(4):345–56.

Helkimo E, Carlsson GE, Helkimo M. Chewing efficiency and state of dentition. Acta Odontol Scand. 1978;36:33–41.

Herschel A, et al. Mechanism of intraoral transport in macaques. Am J Phys Anthropol. 1984;65:275–82.

Hiiemae KM. Feeding in mammals. In: Schwenk K, editor. Feeding: form, function, and evolution in tetrapod vertebrates. San Diego, CA: Academic Press; 2000.

Hiiemae KM, Palmer J. Food transport and bolus formation during complete feeding sequences on foods of different initial consistency. Dysphagia. 1999;14:31–42.

Hiiemae K, et al. Natural bites, food consistency and feeding behaviour in man. Arch Oral Biol. 1996;41(2):175–89.

Hiss SG, Treole K, Stuart A. Effects of age, gender, bolus volume, and trial on swallowing apnea duration and swallow/respiratory phase relationships of normal adults. Dysphagia. 2000;16:128–35.

Hodgson MI, Linforth RS, Taylor AJ. Simultaneous real-time measurements of mastication, swallowing, nasal airflow, and aroma release. J Agric Food Chem. 2003;51(17):5052–7.

Hofmann W, et al. Working memory and self-regulation. In: Vohs KD, Baumeister RF, editors. Self- regulation. Research, theory and applications. New York: The Guilford Press; 2011.

HumbertI A, Robbins J. Dysphagia in the elderly. Physl Med Rehabil Clin North Am. 2008;19:853–66.

Humphrey T. The development of human fetal activity and its relation to postnatal behavior. In: Reese HW, Lipsitt LP, editors. Advances in child development and behavior, Vol, vol. 5; 1970. p. 1–57.

Hylander WL, Johnson KR, Crompton AW. Loading patterns and jaw movements during mastication in macaca fascicularis: a bone-strain, electromyographic, and cineradiographic analysis. Am J Phys Anthropol. 1987;72(3):287–314.

ID. Nose and Eustachian Tube. Roma: CIC Edizioni Internazionali; 1989.

ID. Mental imagery in the motor context. Neuropsychologia. 1995;33:1419–32.

ID. Importanza di una corretta deglutizione. San Benedetto del Tronto: Futura Publishing Society; 1996.

ID. The psychology of attention. Cambridge, MA: The MIT Press; 1998.

ID. Development of working memory: should the Pascual Leone and the Baddeley and hitch models be merged? J Exp Child Psychol. 2000a;77(2):128–37.

ID. Task switching and multitask performance. In: Monsell S, Driver J, editors. Attention and performance xviii: control of mental processes. The MIT Press: Cambridge, MA; 2000b.

ID. Executive attention, working memory capacity and a two-factor theory of cognitive control. In: Ross B, editor. The psychology of learning and motivation: advances in research and theory, vol. 44. New York: Elsevier Science; 2004a. p. 145–99.

ID. I protocolli riabilitativi di tipo cognitivo integrati con trattamenti attentivi: alcune considerazioni teoriche e sperimentali a sostegno. Giornale Ital delle Disabil. 2004b;4(3):41–8.

ID. Tempi di reazione e biochimica dei sistemi attentivi. Approfondimenti. In: Feldman R, editor. Psicologia generale. Milano: McGraw-Hill; 2008a. p. 60–1.

ID. Reti cerebrali e neurofisiologia funzionale dei sistemi attentivi. Approfondimenti. In: Feldman R, editor. Psicologia generale. Milano: McGraw-Hill; 2008b. p. 55–6.

ID. Pulmonary aspiration syndromes. Curr Opin Pulm Med. 2011a;17:148–54.

ID (2011b), Introduzione. In: O. Schindler, G. Ruoppolo, A. Schindler (eds) Deglutologia. Omega: Torino.

ID. Oral phase preparation and propulsion: anatomy, physiology, rheology, mastication, and transport. In: Shaker R, et al., editors. Principles of deglutition. New York: Springer; 2013.

ID. Attenzione esecutiva, memoria e autoregolazione. Una riflessione neuroscientifica su funzionamento, assessment, (ri)abilitazione. Firenze: Hogrefe Editore; 2018.

ID. La semplessità, Ed. Torino: Codice; 2019.

ID. Typical feeding and swallowing development in infants and children. In: Groher ME, Crary MA, editors. Dysphagia, clinical management in adult and children. 3th ed. New York: Elsevier; 2021.

ID. Disturbi del linguaggio e disprassia verbale (a cura di). Roma: Carocci; 2024.

Jansma JM, et al. Functional anatomical correlates of controlled and automatic processing. J Cogn Neurosci. 2001;13(6):730–43.

Jean A. Brain stem control of swallowing: neuronal network and cellular mechanisms. Physiol Rev. 2001;81:929–69.

Jeannerod M. The representing brain: neural correlates of motor intention and imagery. Behav Brain Sci. 1994;17:187–245.

Jonas I, et al. Relationship between tubal function, craniofacial morphology and disorder of deglutition. Arch Oto-Rhino-Laryngol. 1978;2a(3–4):151–62.

Kandel ER, Schawartz JH. Principi di Neuroscienze. Milano: Casa editrice Ambrosiana; 1991.

Kelly AM, Garavan H. Human functional neuroimaging of brain changes associated with practice. Cereb Cortex. 2005;15:1089–102.

Kelly AM, et al. Competition between functional brain networks mediates behavioral variability. NeuroImage. 2008;39(1):527–37.

Kendall KA, Louie S. Severe obstructive airway disorders and diseases: vocal fold dysfunction. Clin Rev Allergy Immunol. 2003;25(3):221–31.

Kennedy JG, Kent RD. Physiological substrates of normal deglutition. Dysphagia. 1988;3(1):24–37. (3rd ed.)

Kenner C, McGrath JM. Developmental care of newborns and infants: a guide for health professionals. St. Louis, MO: Mosby; 2010.

Kliegman R, et al. Nelson textbook of pediatrics. 19th ed. Philadelphia, PA: Elsevier; 2011.

Kossioni AE, Dontas AS. The stomatognathic system in the elderly. Useful information for the medical practitioner. Clin Interv Aging. 2007;2(4):591–7.

Kuhn T. La tensione essenziale. Cambiamenti e continuitiz nella scienza. Torino: Einaudi; 1985.

Lama D, Goleman D. Emozioni distruttive. Liberarsi dai tre veleni della mente: rabbia, desiderio e illusione. Milano: Mondadori; 2009.

Landau SM, et al. A functional MRI study of the influence of practice on component processes of working memory. Neuroimage. 2004;22(1):211–21.

Landi F, et al. Sarcopenia: an overview on current definitions, diagnosis and treatment. Curr Protein Pept Sci. 2018;19:1–6.

Lanteri C, Lanteri V. Elementi di ortodonzia. In: Schindler O, Ruoppolo G, Schindler A, editors. Deglutologia. Torino: Omega; 2001.

Lanteri C, Lanteri V, Beretta M. Elementi di ortodonzia. In: Schindler O, Ruoppolo G, Schindler A, editors. Deglutologia. Torino: Omega; 2011.

Le Jemtel TH, Padeletti M, Jelic S. Diagnostic and therapeutic challenges in patients with coexistent chronic obstructive pulmonary disease and chronic heart failure. J Am Coll Cardiol. 2007;49(2):s171–80.

Leisman G, Braun-Benjamin O, Melillo R. Cognitive-motor interactions of the basal ganglia in development. Front Syst Neurosci. 2014; https://doi.org/10.3389/fnsys.2014.00016.

Leisman G, Moustafa AA, Shafir T. Thinking, walking, talking: integratory motor and cognitive brain function. Front Public Health J. 2016;4:94.

Levrini L. Premessa. In: Levrini L, editor. Terapia miofunzionale orofacciale. EDRA: Milano; 2020.

Liu RP, et al. Salivary flow rates in patients with head and neck cancer 0.5 to 25 years after radiotherapy. Oral Surg Oral Med Oral Pathol. 1990;70(6):724–79.

Llinás RR. I of the vortex: from neurons to self. Cambridge (MA): The MIT Press; 2000.

Logemann JA, et al. The evaluation and treatment of swallowing disorders. Curr Opin Otolaryngol Head Neck Surg. 1998;6(6):395–400.

Logemann JA, et al. Temporal and biomechanical characteristics of oropharyngeal swallow in younger and older men. J Speech Lang Hear. 2000;43(1–5):1264–74.

Logemann JA, et al. Oropharyngeal swallow in older and younger women: videofluoroscopic analysis. J Speech Lang Hear. 2002;45:434–45.

Luria AR. Come lavora il cervello. In: Introduzione alla neuropsicologia. Bologna: il Mulino; 1977.

Madhavan A, et al. Prevalence of and risk factors for dysphagia in the community dwelling elderly: a systematic review. J Nutr Health Aging. 2016;20(8):806–15.

Malinowski P. Neural mechanisms of attentional control in mindfulness meditation. Front Neurosci. 2013;7(7):8.

Manabe T, et al. Risk factors for aspiration pneumonia in older adults. PLoS One. 2015; https://doi.org/10.1371/journal.pone.0140060.

Mancuso V. Questa vita. Milano: Garzanti; 2015.

Mandolesi L. Neuroscienza dell'attività motoria. Milano: Springer Italia; 2012.

Marik PE. Aspiration pneumonitis and aspiration pneumonia. N Engl J Med. 2001;344(9):665–71.

Marrocco RT, Davidson MC. Neurochemistry of attention. The attentive brain. Cambridge, MA: The MIT Press; 2000.

Martin BJ, et al. Coordination between respiration and swallowing: respiratory phase relationships and temporal integration. J Appl Physiol. 1994a;76(2):714–23.

Martin BJ, et al. The frequency of respiration and deglutition: influence of posture and oral stimuli. Dysphagia. 1994b;9:78.

Martin-Harris B. Optimal patterns of care in patients with chronic obstructive pulmonary disease. Semin Speech Lang. 2000;21(4):311–21. quiz 320-1

Martin-Harris B, et al. Temporal coordination of pharyngeal and laryngeal dynamics with breathing during swallowing: single liquid swallows. J Appl Phisiol. 2003;94(5):1735–43.

Maspero C, et al. Atipical swallowing: a review. Minerva Stomatol. 2014;63(6):217–27.

Matsuo K, Palmer JB. Anatomy and physiology of feeding and swallowing: normal and abnormal. Phys Med Rehabil Clin North Am. 2008;19(4):691–707.

Matsuo K, Hiiemae KM, Palmer JB. Cyclic motion of the soft palate in feeding. J Dent Res. 2005;84(1):39–42.

Matsuo K, et al. Effects of respiration on soft palate movement in feeding. J Dent Res. 2010;89(12):1401–6.

McCabe DP. The relationship between working memory capacity and executive functioning: evidence for a common executive attention construct. Neuropsychology. 2010;24(2):222–43.

McConnel FM, Cerenko D, Mendelsohn M. Manofluorographic analysis of swallowing. Otolaryngol Clin North Am. 1988;21(4):625–35.

Melillo R, Leisman G. Why the brain works the way it does: evolution and cognition from movement. In: ID, editor. Neurobehavioral disorders of childhood. Springer: New York; 2009.

Menon V. Salience network. In: Brain mapping: an encyclopaedic reference. Stanford: Stanford University School of Medicine; 2015.

Menon V, Uddin LQ. Saliency, switching, attention and control: a network model of insula function. Brain Struct Funct. 2010;214(5-6):655–67.

Metzler-baddeley C, et al. Task complexity and location specific changes of cortical thickness in executive and salience networks after working memory training. Neuroimage. 2016;130:48–62.

Metzler-baddeley C, et al. Dynamics of white matter plasticity underlying working memory training: multimodal evidence from diffusion mRu and relaxometry. J Cogn Neurosci. 2018;29(9):1509–20.

Middleton FA, Strick PL. Basal ganglia output and cognition: evidence from anatomical, behavioral, and clinical studies. Brain Cogn. 2000;42(2):183–200.

Miller JL, Sonies BC, Macedonia C. Emergence of oropharyngeal, laryngeal and swallowing activity in the developing fetal upper aerodigestive tract: an ultrasound evaluation. Early Human Dev. 2003;71(L):61–87.

Mokhlesi B, et al. Oropharyngeal deglutition in stable COPD. Chest. 2002;121(2):361–9.

Morgon A, Nancy PH, Deyean Y. L'otite séro-muqueuse et ses complications. Paris: Librerie Arnette; 1985.

Morris SE, Klein MD. Pre-feeding skills: a comprehensive resources for feeding development. Austin, TX: Pro-Ed; 2000.

Moruzzi G, Magoun HW. Brain stem reticular formation and activation of the EEG. Electroencephalogr Clin Neurophysiol. 1949;1(4):455–73.

Mosier KM, et al. Lateralization of cortical function in swallowing: a functional MR imaging study. Am J Neuroradiol. 1999;20(8):1520–6.

Narhi TO. Prevalence of subjective feelings of dry mouth in the elderly. J Dent Res. 1994;73:20–5.

Ney DM, et al. Senescent swallowing: impact, strategies, and interventions. Nutr Clin Pract. 2009;24:395–413.

Nicosia MA, et al. Age effects on the temporal evolution of isometric and swallowing pressure. J Gerontol Ser A Biol Sci Med Sci. 2000;55(11):M634–40.

Olesen PJ, Westerberg H, Klingberg T. Increased prefrontal and parietal activity after training of working memory. Nat Neurosci. 2004;7(1):75.

Osterberg T, Carlsson GE. Symptoms and signs of mandibular dysfunction in 70-year-old men and women in Gothenburg, Sweden. Commun Dent Oral Epidemiol. 1979;7(6):315–21.

Palmer JB, Hiiemae KM. Eating and breating: interaction between respiration and feeding on solid food. Dysphagia. 2003;18(3):169–78.

Palmer JB, et al. Coordination of mastication and swallowing. Dysphagia. 1992;7:187–200.

Parisi G. In un volo di storni. Milano: Rizzoli; 2021.

Pashler H. Dual-task interference in simple tasks: data and theory. Psychol Bull J. 1994;116(2):220–44.

Passali D. L'unità rino-faringo tubarica. CRS Amplifon: Milano; 1985.

Patel P, Lamar M, Bhatt T. Effect of type of cognitive task and walking speed on cognitive-motor interference during dual-task walking. Neuroscience. 2014;260:140–8.

Paydarfar D, Eldridge FL, Kiley JP. Resetting of mammalian respiratory rhythm: existence of a phase singularity. Am J Physiol. 1986;250(4 (Pt. 2)):R721–7.

Percival RS, Challacombe SJ, Marsh PD. Flow rates of resting whole and stimulated parotid saliva in relation to age and gender. J Dent Res. 1994;73(8):1416–20.

Perlman AL, et al. Bolus location associated with videofluoroscopic and respirodeglutometric events. J Speech Lang Hear Res. 2005;48(1):21–33.

Petersen SE, Posner MI. The attention system of the human brain: 20 years after. Annu Rev Neurosci. 2012;35:73–89.

Petrosino L, Fucci D, Robey RR. Changes in lingual sensitivity as a function of age and stimulus exposure time. Perceptual Motor Skills. 1982;55(3(Pt. 2)):1083–90.

Poldrack RA. Can cognitive processes be inferred from neuroimaging data? Trends Cogn Sci. 2006;10(2):59–63.

Polimeni A. Presentazione. In: Levrini L, editor. Terapia miofunzionale orofacciale. EDRA: Milano; 2020.

Pribram K. H. (1973), The primate frontal cortex-executive of the brain, in K. H. Pribram, A. R. Lurija (eds.), Psychophysiology of the frontal lobes, Academic Press, New York, pp. 293–314.

Proffit WR. Ortodonzia moderna. 2nd ed. Elsevier Masson: Milano; 2001.

Proffit WR, Mason RM. Myofunctional therapy for tongue-thrusting: background and recommendations. J Am Dent Assoc. 1975;90(2):403–n.

Prosiegel M, et al. Kinematic analysis of laryngeal movements in patients with neurogenic dysphagia before and after swallowing rehabilitation. Dysphagia. 2000;15(4):173–9.

Rabaglietti E, De Lorenzo A, Brustio PR. The role of working memory on dual-task cost during walking performance in childhood. Front Psychol. 2019;10.

Rabbitt P. Methodologies and models in the study of executive function. In: Rabbitt P, editor. Methodology of frontal and executive function. Howe, TX: Psychology Press; 1997. p. 1–38.

Raichle ME, et al. A default mode of brain function. Proc Natl Acad Sci U S A. 2001;98(2):676–82.

Repovš G, Baddeley A. The multi-component model of working memory: explorations in experimental cognitive psychology. Neuroscience. 2006;139(1):5–21.

Riu R, Flottes L, Bouche J. La physiologie de la Trompe d'Eustache: applications cliniques et thérapeutiques. Paris: Librerie Arnette; 1966.

Rix RF. Deglutition and the teeth. Dent Record. 1946;66:103–8.

Rizzolatti G, Sinigaglia C. So quel che fai. Milano: Raffaello Cortina Editore; 2006.

Robbins J, et al. Oropharyngeal swallowing in normal adults of different ages. Gastroenterology. 1992;103(3):823–9.

Robbins J, et al. Age effects on lingual pressure generation as a risk factor for dysphagia. J Gerontol Ser A Biol Sci Med Sci. 1995;50(5):M257–62.

Rofes L, et al. Pathophysiology of oropharyngeal dysphagia in the frail elderly. Neurogastroenterol Motil. 2010;22:851–8. e230

Rogers JH. Swallowing patterns of a normal-population sample compared to those of patients from an orthodontic practice. Am J Orthod Dentofac Orthop. 1961;17:674–9.

Rosenfeld RM, et al. Clinical practice guideline: otitis media with effusion (update). Otolaryngol Head Neck Surg. 2016;154(1 Suppl):S1–S41.

Rossi G, et al. Reliability of the Kinovea bidimensional system for measuring buccal movement in a sample of Italian adults. In: Proceedings of fifth international congress on information and communication technology, vol. 2. London: ICICT; 2020. p. 184–93.

Sabbadini L. Disturbi specifici del linguaggio, disprassie e funzioni esecutive. Milano: Springer; 2013.

Sakai K, Rowe JB, Passingham RE. Active maintenance in prefrontal area 46 creates distractor-resistant memory. Nat Neurosci. 2002;5(5):479.

Schindler O. Manuale operativo di fisiopatologia della deglutizione. Torino: Omega; 1990.

Schindler A. Presentazione. In: Levrini L, editor. Terapia miofunzionale orofacciale. Milano: EDRA; 2020.

Schindler A, et al. Fisiologia della deglutizione. In: Schindler O, Ruoppolo G, Schindler A, editors. Deglutologia. Torino: Omega; 2011.

Schmidt RA. Motor control and learning: a behavioural emphasis. Champaign, IL: Human Kinetics; 1982.

Schmidt RA, Wrisberg CA. Motor learning and performance. Champaign, IL: Human Kinetics; 2008.

Schwaighofer M, Fisher F, Buhner M. Does working memory training transfer? A meta-analysis including training conditions as moderators. Educ Psychol. 2015;50:138–66.

Seidler RD, Bo J, Anguera JA. Neurocognitive contributions to motor skill learning: the role of working memory. J Mot Behav. 2012;44(6):445–53.

Serra A. Terapia medica delle ipoacusie trasmissive. In: Paludetti G, editor. Ipoacusie infantili. Dalla diagnosi alla terapia. Torino: Omega; 2011.

Shaker R, Lang IM. Effect of aging on the deglutitive oral, pharyngeal, and esophageal motor function. Dysphagia. 1994;9(4):221–8.

Shaker R, et al. Coordination of deglutition and phases of respiration: effect of aging, tachypnea, bolus volume, and chronic obstructive pulmonary disease. Am J Physiol. 1992;263(5 (Pt. 1)):G750–5.

Shaker R, et al. Effect of aging, position, and temperature on the threshold volume triggering pharyngeal swallows. Gastroenterology. 1994;107:396–402.

Shern RJ, Fox PC, Li SH. Influence of age on the secretory rates of the human minor salivary glands and whole saliva. Arch Oral Biol. 1993;38(9):755–61.

Ship JA. The influence of aging on oral health and consequences for taste and smell. Physiol Behav. 1999;66:209–15.

Smithard DG. Dysphagia: a geriatric giant? Med Clin Rev. 2016;2:5.

Speidel TM, Isaacson RJ, Worms FW. Tongue-thrust therapy and anterior dental open-bite. A review of new facial growth data. Am J Orthod. 1972;62(3):287–95.

Sridharan D, Levitin DJ, Menon V. A critical role for the right fronto-insular cortex in switching between central-executive and default-mode networks. Proc Natl Acad Sci. 2008;105(34):12569–74.

Stahl F, et al. Relationship between occlusal findings and orofacial myofunctional status in primary and mixed dentition. Part II: prevalence of orofacial dysfunctions. J Orofac Orthop. 2007;68(2):74–90.

Stephen JR, et al. Bolus location at the initiation of the pharyngeal stage of swallowing in healthy older adults. Dysphagia. 2005;20(4):266–72.

Stevens JC, Cain WS. Age-related deficiency in the perceived strength of six odorants. Chem Sens. 1985;10(4):517–29.

Stiles J, Jernigan TL. The basics of brain development. Neuropsychol Rev. 2010;20(4):327–48.

Straub WJ. Malfunction of the tongue. Part I. The abnormal swallowing habit: its cause, effects, and results in relation to orthodontic treatment and speech therapy. Am J Orthod. 1960;46(6):404–24.

Takeuchi H. Training of working memory impacts structural connectivity. J Neurosci. 2010;30:3297–303.

Takeuchi H, Taki Y, Kawashima R. Effects of working memory training on cognitive functions and neural systems. Rev Neurosci. 2010;21:427–49.

Tang YY, Rothbart MK, Posner MI. Neural correlates of establishing, maintaining, and switching brain states. Trends Cogn Sci. 2012;16(6):330–7.

Tara L, Lamberg EM, Muratori LM. Building a framework for a dual task taxonomy. Biomed Res Int. 2015;9:10.

Teismann IK, et al. Age-related changes in cortical swallowing processing. Neurobiol Aging. 2010;31:1044–50.

Thexton AJ, Hiiemae KM. The effect of food consistency upon jaw movement in the macaque: a cineradiographic study. J Dent Res. 1997;76:552–60.

Turatto M. Change Blindness: guardare senza vedere: Una nuova prospettiva nello studio dell'attenzione visiva. G Ital Psicol. 2000;4:679–700.

Turatto M, et al. Non-spatial attentional shifts between audition and vision. J Exp Psychol Human Perform Percept. 2002;28:628–39.

Uddin LQ. Salience processing and insular cortical function and dysfunction. Nat Rev Neurosci. 2014;16:55–61.

Uddin LQ, et al. Functional connectivity of default mode network components: correlation, anti-correlation, and causality. Hum Brain Mapp. 2009;30:625–37.

Varela-Vásquez LA, Minobes-Molina E, Jerez-Roig J. Dual-task exercises in older adults: a structured review of current literature. J Frailty Sarcopenia Falls. 2020;5(2):31–7.

Veneroso MC, et al. Dalla teoria alla pratica: un progetto di didattica integrata. Annali online della Didattica e della Formazione Docenti. 2016;8(11):123–33.

Vogt S, Magnussen S. Expertise in pictorial perception: eye-movement patterns and visual memory in artists and laymen. Brain Cogn. 2007;58(3):324–33.

Vroon R, et al. Il seduttore segreto, psicologia dell'olfatto. Roma: Editori Riuniti; 2003.

Wagner AD, et al. Building memories: remembering and forgetting of verbal experiences as predicted by Brain activity. Science. 1998;281:1188–91.

Werlich E. The Prevalence of Variant Swallowing Patterns in a Group of Seattle School Children, Master's thesis, University of Washington; 1962.

Wolf LS, Glass RR. Feeding and swallowing disorders in infancy: assessment and management, therapy skill builders. Tucson, AZ; 1992.

Zheng Z, et al. Silent aspiration in patients with exacerbation of COPD. Eur Respir J. 2016;8(2):570–3.